Numerical Integration of Stochastic Differential Equations

Mathematics and Its Applications

Volume 313

Numerical Integration of Stochastic Differential Equations

by

G.N. Milstein

Department of Mathematics,
Ural State University,
Ekatarinburg, Russia

KLUWER ACADEMIC PUBLISHERS

DORDRECHT / BOSTON / LONDON

Library of Congress Cataloging-in-Publication Data

Mil'shteĭn, G. N. (Grigoriĭ Noĭkhovich)
[Chislennoe integrirovanie stokhasticheskikh differentsial'nykh uravneniĭ. English]
Numerical integration of stochastic differential equations / by G.N. Milstein.
p. cm. -- (Mathematics and its applications ; v. 313)
Includes bibliographical references and index.

1. Stochastic differential equations--Numerical solutions. 2. Wiener integrals. I. Title. II. Series: Mathematics and its applications (Kluwer Academic Publishers) ; v. 313.
QA274.23.M5513 1995
519.2--dc20 94-37674

ISBN 978-90-481-4487-7

Published by Kluwer Academic Publishers,
P.O. Box 17, 3300 AA Dordrecht, The Netherlands.

Kluwer Academic Publishers incorporates
the publishing programmes of
D. Reidel, Martinus Nijhoff, Dr W. Junk and MTP Press.

Sold and distributed in the U.S.A. and Canada
by Kluwer Academic Publishers,
101 Philip Drive, Norwell, MA 02061, U.S.A.

In all other countries, sold and distributed
by Kluwer Academic Publishers Group,
P.O. Box 322, 3300 AH Dordrecht, The Netherlands.

This is a revised and updated translation of the original Russian work
Numerical Integration of Stochastic Differential Equations,
Ural State University Press, Sverdlovsk © 1988

Printed on acid-free paper

Contents

Introduction

Using stochastic differential equations we can successfully model systems that function in the presence of random perturbations. Such systems are among the basic objects of modern control theory. However, the very importance acquired by stochastic differential equations lies, to a large extent, in the strong connections they have with the equations of mathematical physics. It is well known that problems in mathematical physics involve 'damned dimensions', often leading to severe difficulties in solving boundary value problems. A way out is provided by stochastic equations, the solutions of which often come about as characteristics.

In its simplest form, the *method of characteristics* is as follows. Consider a system of n ordinary differential equations

$$dX = a(X)\,dt. \tag{0.1}$$

Let $X_x(t)$ be the solution of this system satisfying the initial condition $X_x(0) = x$. For an arbitrary continuously differentiable function $u(x)$ we then have:

$$u(X_x(t)) - u(x) = \int_0^t \left(a(X_x(t)), \frac{\partial u}{\partial x}(X_x(t)) \right) dt. \tag{0.2}$$

We consider the Cauchy problem for the first-order linear partial differential equation

$$\left(a(x), \frac{\partial u}{\partial x} \right) = 0, \tag{0.3}$$

$$u|_\gamma = f(x), \tag{0.4}$$

where γ is a curve in the n-dimensional space of the variable x. Let u be a solution of equation (0.3). Then (0.2) implies

$$u(x) = u(X_x(t)). \tag{0.5}$$

Formula (0.5) indicates the following way for solving the problem (0.3)–(0.4): starting at x, draw the trajectory $X_x(t)$ of the system (0.1) up to the moment τ of its intersection with γ. By (0.4), u is known on γ. Therefore

$$u(x) = u(X_x(\tau)) = f(X_x(\tau)). \tag{0.6}$$

We now consider the system of stochastic differential equations

$$dX = a(X)\,dt + \sum_{r=1}^{q} \sigma_r(X)\,dw_r(t). \tag{0.7}$$

By Itô's formula we obtain for a sufficiently smooth function $u(x)$ the following analog of (0.2):

$$u(X_x(\tau)) - u(x) = \int_0^\tau Lu(X_x(t))\,dt + \sum_{r=1}^{q} \int_0^\tau \Lambda_r u(X_x(t))\,dw_r(t). \tag{0.8}$$

In this formula, τ is a Markov moment, and

$$L = \left(a, \frac{\partial}{\partial x}\right) + \frac{1}{2}\sum_{r=1}^{q}\left(\sigma_r, \frac{\partial}{\partial x}\right)^2 = \sum_{i=1}^{n} a^i \frac{\partial}{\partial x^i} + \frac{1}{2}\sum_{r=1}^{q}\sum_{i,j=1}^{n} \sigma_r^i \sigma_r^j \frac{\partial^2}{\partial x^i \partial x^j},$$

$$\Lambda_r = \left(\sigma_r, \frac{\partial}{\partial x}\right) = \sum_{i=1}^{n} \sigma_r^i \frac{\partial}{\partial x^i},$$

where a^i, σ_r^i are the components of the vectors a, σ_r.

For an elliptic-type equation

$$Lu = 0 \tag{0.9}$$

we consider, in a domain $\mathcal{D}$ with boundary Γ, the *Dirichlet problem* with boundary condition

$$u|_\Gamma = f(x). \tag{0.10}$$

Let u be a solution of (0.9). Then (0.8) implies

$$u(x) = u(X_x(\tau)) - \sum_{r=1}^{q} \int_0^\tau \Lambda_r u(X_x(t))\,dw_r(t). \tag{0.11}$$

Taking for τ the time at which the trajectory $X_x(t)$ hits the boundary Γ, we arrive at a probabilistic representation of the solution of (0.9)–(0.10), thanks to the average in (0.11):

$$u(x) = \mathbf{E}u(X_x(\tau)) = \mathbf{E}f(X_x(\tau)). \tag{0.12}$$

Using the *Monte-Carlo approach* we obtain

$$u(x) = \frac{1}{N}\sum_{m=1}^{N} f(X_x^{(m)}(\tau^{(m)})), \tag{0.13}$$

where $X_x^{(m)}(t)$, $m = 1, \ldots, N$, are independent realisations of the process $X_x(t)$ defined by the system (0.7).

Thus, the multi-dimensional boundary value problem (0.9)–(0.10) reduces to the Cauchy problem for the system (0.7). This system can be naturally regarded as one-dimensional, since it contains only one independent variable. The system (0.7) comes about as characteristic system of differential equations for the problem (0.9)–(0.10). This approach, while enabling reduction of a multi-dimensional boundary value problem to a one-dimensional Cauchy problem, cannot be of importance for numerical mathematics.

We will now give another well-known probabilistic representation of the solution of the *Cauchy problem for the heat equation*

$$Lu \equiv \frac{\partial u}{\partial t} + \sum_{i=1}^{n} a^i(t,x)\frac{\partial u}{\partial x^i} + \frac{1}{2}\sum_{r=1}^{q}\sum_{i=1}^{n}\sum_{j=1}^{n}\sigma_r^i(t,x)\sigma_r^j(t,x)\frac{\partial^2 u}{\partial x^i \partial x^j} = 0, \quad (0.14)$$

$$u(t_0+T,x) = f(x), \quad (0.15)$$

where $t_0 \le t \le t_0+T$, $x \in \mathbb{R}^n$.

The value of the unknown function u at a point (s,x) can be expressed as a mathematical expectation:

$$u(s,x) = \mathbf{E} f(X_{s,x}(t_0+T)), \quad (0.16)$$

where $X_{s,x}(t)$ is the solution of the following system of stochastic differential equations (which is not autonomous, in distinction to (0.7)):

$$\begin{aligned} dX &= a(t,X)\,dt + \sum_{r=1}^{q}\sigma_r(t,X)\,dw_r(t), \\ X_{s,x}(s) &= x, \qquad s \le t \le t_0+T. \end{aligned} \quad (0.17)$$

Application of the Monte-Carlo method gives

$$u(s,x) = \mathbf{E} f(X_{s,x}(t_0+T)) \doteq \frac{1}{N}\sum_{m=1}^{N} f(X_{s,x}^{(m)}(t_0+T)), \quad (0.18)$$

where $X_{s,x}^{(m)}(t_0+T)$, $m = 1,\dots,N$, are independent realisations of the random variable $X_{s,x}(t_0+T)$.

To be able to use (0.18) (see also (0.13)) we have to model the random variable $X_{s,x}(t_0+T)$. The precise computation of $X_{s,x}(t_0+T)$ is impossible even in the deterministic situation. Therefore we have to replace $X_{s,x}(t_0+T)$ by a nearby random variable $\overline{X}_{s,x}(t_0+T)$ that can be modeled. Instead of (0.18) we obtain

$$u(s,x) = \mathbf{E} f(X_{s,x}(t_0+T)) \doteq \mathbf{E} f(\overline{X}_{s,x}(t_0+T)) \doteq \frac{1}{N}\sum_{m=1}^{N} f(\overline{X}_{s,x}^{(m)}(t_0+T)). \quad (0.19)$$

The first approximate equality in (0.19) involves an error brought about by replacing X by $\overline{X}$ (an error related with the approximate integration of the system (0.17)); in the second approximate equality the error comes from the Monte-Carlo method. Hence arises the pressing need for creating numerical integration methods for stochastic equations. The theory of these methods has only just begun to develop.

We note that there are many papers (see [14], [17], and the references in these) in which, in order to use the Monte-Carlo method, probabilistic representations are chosen that contain, in distinction to the representation (0.16), 'realisable' (see [17]) random variables. However, using a representation of the form (0.16) in the spirit of (0.19) does, without doubt, have its merits.

We partition the interval $[t_0, t_0+T]$ into N equal parts using division points t_k, so that $t_{k+1}-t_k=h$, $k=0,1,\dots,N-1$, $t_0+T=t_N$, $h=T/N$. We will denote the approximation to $X(t_k)$ by $\overline{X}(t_k)$, $\overline{X}_k$, or simply X_k. Everywhere below we have put $\overline{X}_0 = X(t_0)$.

The simplest approximate method for solving (0.17) is *Euler's method*:

$$X_{k+1} = X_k + \sum_{r=1}^{q} \sigma_{r_k} \Delta_k w_r(h) + a_k h. \tag{0.20}$$

In (0.20), $\Delta_k w_r(h) = w_r(t_{k+1}) - w_r(t_k)$, and the index k at σ_r and a indicates that these functions are evaluated at the point (t_k, X_k).

G. Marujama [52] has shown the mean-square convergence of this method, while I.I. Gikhman and A.V. Skorokhod [8], [10] have shown that the order of accuracy of Euler's method is 1/2, i.e.

$$\left[\mathbf{E}(X(t_k)-X_k)^2\right]^{1/2} \le C h^{1/2}, \tag{0.21}$$

where C is a constant not depending on k and h (but, clearly, depending on the system (0.7) and $X(t_0)$).

If for some method we would have

$$\left[\mathbf{E}(X(t_k)-X_k)^2\right]^{1/2} \le C h^{p} \tag{0.22}$$

where C does not depend on k and h and where $p>0$, then we say that the *mean-square order of accuracy* of the method is p.

A method of second order of accuracy has first been constructed in [25]. It is as follows:

$$X_{k+1} = X_k + \sum_{r=1}^{q} \sigma_{r_k} \Delta_k w_r + a_k h + \sum_{i=1}^{q}\sum_{r=1}^{q} (\Lambda_i \sigma_r)_k \int_{t_k}^{t_{k+1}} (w_i(\theta) - w_i(t_k))\, dw_r(\theta), \tag{0.23}$$

where $\Lambda_i = \left(\sigma_i, \frac{\partial}{\partial x}\right)$.

For a single noise ($q=1$) the integral in (0.23) can be expressed in terms of $\Delta_k w$, and the formula takes the following form, e.g. in the scalar case:

$$X_{k+1} = X_k + \sigma_k \Delta_k w + a_k h + \frac{1}{2}\left(\sigma\frac{\partial\sigma}{\partial x}\right)_k \Delta_k^2 w - \frac{1}{2}\left(\sigma\frac{\partial\sigma}{\partial x}\right)_k h. \tag{0.24}$$

The same paper contains methods of higher order of accuracy for the scalar case.

N.J. Rao, J.D. Borwankar, and D. Ramkrishna have considered scalar equations in [55], and convergence of the methods could be proved only in the sense of convergence in probability. They also obtained (0.24). N.N. Nikitin and V.D. Razevig [34], W. Rumelin [56], D. Talay [57]–[60], E. Pardoux and D. Talay [54], and others have obtained results that border on those of [25], [55].

We must note that the numerical integration problem under consideration is close to the *estimation problem*, given some information on the Wiener process participating in (0.17).

For example, suppose we know only the *increments* $\Delta_m w_r(h)$, $r \doteq 1, \dots, q$, $m = 0, 1, \dots, k-1$ of the Wiener process, and suppose it is required to find approximations X_k to $X(t_k)$. It is obvious that the estimator

$$\widehat{X}(t_k) = \mathbf{E}\{X(t_k) \mid \Delta_m w_r(h),\ r = 1, \dots, q,\ m = 0, \dots, k-1\} \tag{0.25}$$

is best in the sense of mean-square. Such estimators have been constructed in [31] for systems of linear differential equations with additive noises:

$$dX = A(t)X\,dt + \sum_{r=1}^{q} \sigma_r(t)\,dw_r(t). \tag{0.26}$$

In the same paper, for systems of the form (0.26) there has been constructed a numerical integration method that uses the modeling discrete processes $\Delta_k w_r$ in an optimal manner. Already for the scalar equation

$$dX = aX\,dt + \sigma\,dw$$

with constant coefficients $a \neq 0$, $\sigma \neq 0$ this method is of precisely first order of accuracy (note that for (0.26) even Euler's method is of first order of accuracy, but in a number of cases an optimal method gives essentially more refined results [31]). Thus, there is no numerical integration method for (0.26) that uses only information about $w_r(t)$, $r = 1, \dots, q$, at discrete moments of time t_k and would have order of accuracy exceeding $O(h)$ (see also J.M.C. Clark and R.T. Cameron [47] for this).

In the case of a general system (0.17), using only information about $w_r(t)$, $r = 1, \dots, q$, at discrete moments of time t_k we can only construct a method of order of accuracy 1/2. Methods of higher order of accuracy have been constructed in [25]; these have to use additional information regarding $w_r(t)$. For example, the method (0.23) uses at each step the integrals $\int_{t_k}^{t_{k+1}} (w_i(\theta) - w_i(t_k))\,dw_r(\theta)$, $i, r = 1, \dots, q$. In the case of a single noise, invoking the integrals $\int_{t_k}^{t_{k+1}} (w(\theta) - w(t_k))\,d\theta$ allows us to construct a method of order 3/2, while additionally invoking the integrals $\int_{t_k}^{t_{k+1}} (w(\theta) - w(t_k))^2\,d\theta$ leads to a method of order 2 [25].

The methods by which a number of formulas have been obtained in [25] are somewhat laborious. They use the theory of Markov operator semigroups in combination with the method of undetermined coefficients.

W. Wagner and E. Platen [62] have given a very simple proof (using only Itô's formula) of the expansion of the solution $X_{t,x}(t+h)$ of the system (0.17) in powers of h and in integrals depending on the increments $w_r(\theta) - w_r(t)$, $t \leq \theta \leq t+h$, $r = 1, \dots, q$. This expansion generalises (0.23). In the deterministic situation it comes down to Taylor's formula for $X_{t,x}(t+h)$ in powers of h in a neighborhood of the point (t, x). Theoretically, this expansion allows one to construct methods of arbitrarily high order of accuracy (of course, with corresponding conditions about the coefficients of the system (0.17)). We recall that in the deterministic situation, methods based on direct expansion of the solution by Taylor's formula have not at all become widespread.

This is first of all related to the problem of computing derivatives of the righthand side of the system. To solve such problems for ordinary differential equations one has created *methods of Runge–Kutta type* and multistep *difference methods*. Some results in this direction for stochastic differential equations have led to the present book. In it we give a construction of implicit numerical integration methods for stochastic equations (as is well known, *implicit methods* turned out to be necessary for the numerical integration of *stiff systems* of ordinary differential equations, see, e.g., [39], [43]).

Above we have already noted that for the construction of methods of higher order of accuracy it is necessary to invoke more general information about the Wiener processes $w_r(t)$, $r = 1, \dots, q$. In relation with this there arises the inevitable problem of modeling complicated random quantities. For example, in the mean-square method (0.23) of first order of accuracy it was necessary to approximately model the *Itô integrals*

$$\int_0^h w_i(s)\, dw_j(s), \qquad i, j = 1, \dots, q.$$

While modeling of the solution of a system of stochastic differential equations is a prerequisite for using the Monte-Carlo method (see, e.g., formula (0.19) in relation to the problem (0.14)–(0.15)), it is not at all necessary to solve the very complicated problem of finding mean-square approximations. Let $X(t)$ be the exact and $\overline{X}(t)$ an approximate solution. In many problems of mathematical physics it is only required that the mathematical expectation $\mathbf{E}f(\overline{X}(t))$ be close to $\mathbf{E}f(X(t))$ for a sufficiently large class of functions f, i.e. that $\overline{X}(t)$ be close to $X(t)$ in a weak sense. Suppose we can solve the problem of computing $\mathbf{E}f(X(t_0 + T))$. If an approximation $\overline{X}$ is such that

$$\left|\mathbf{E}f(\overline{X}(t_0 + T)) - \mathbf{E}f(X(t_0 + T))\right| \le C\, h^p \tag{0.27}$$

for f from a sufficiently large class of functions, then we say that the *weak order of accuracy* (or, if this does not cause confusion, simply the *order of accuracy*) of the approximation $\overline{X}$ (the method $\overline{X}$) is p. We can prove, as an example, that the weak order of accuracy of Euler's method is 1. Note that for numerical integration in the mean-square sense with some order of accuracy we can guarantee an approximation in the weak sense with the same order of accuracy, since if $(\mathbf{E}|\overline{X}(t) - X(t)|^2)^{1/2} = O(h^p)$, then for every function f satisfying a Lipschitz condition we have $\mathbf{E}(f(\overline{X}(t)) - f(X(t))) = O(h^p)$. Moreover, an increase in the order of accuracy in the mean-square sense does not, in general, imply an increase of the weak order of accuracy. For example, the method (0.23) has first order of accuracy, as has Euler's method. At the same time, a 'crude' method like (we give the formula for a scalar equation):

$$X_{k+1} = X_k + a_k h + \sigma_k \alpha_k h^{1/2}, \tag{0.28}$$

where α_k, $k = 0, 1, \dots, N-1$, are independent random variables taking the values $+1$ and -1 with probabilities $1/2$, also has first order of accuracy in the sense of weak approximation. Furthermore, it has been shown in [26] that, using the modeling $\Delta_k w(h) = w(t_k + h) - w(t_k)$ (and also when modeling by simpler random variables), for scalar equations of general kind we can construct a method of second order of accuracy in the weak sense, while at the same time we cannot construct in this way a method of order of accuracy exceeding one in the mean-square sense, not even for the linear equation $dX = aX\,dt + \sigma\,dw$ with constant coefficients $a \neq 0$, $\sigma \neq 0$.

All this testifies of the fact that both *mean-square approximations* and *weak approximations* are of independent interest. As far as known to the author, weak approximations have been introduced for the first time ever in [26]. The main interest in weak approximations lies in the hope to obtain simpler methods and, in particular, methods not requiring modeling by complicated random variables. We recall that, e.g., in the mean-square method [23] already for the first order accuracy there arises the nonsimple problem of modeling random variables of the type $\int_0^h w_i(\theta)\,dw_j(\theta)$. These problems of modeling complicated random variables can be avoided by integrating in the weak sense, which is an impetus for the development of methods for constructing weak approximations.

In [28] both implicit and explicit constructive (from the point of view of modeling random variables) numerical integration methods (in the weak sense) have been given that are of second order of accuracy for general systems of stochastic differential equations; in it are also given methods of third order of accuracy for systems with additive noises. We note that while in the deterministic theory the one-dimensional case differs but little from the multi-dimensional one, for the numerical integration of stochastic differential equations the multi-dimensional case, especially when several noises are involved, is essentially more complicated than the one-dimensional case.

Weak approximations have also been considered by E. Pardoux and D. Talay in [54]. In that article, and also in D. Talay's work [57], [58], trajectory-wise approximation is considered, while in [55] approximation in the sense of convergence in probability is considered. These kinds of approximation are of far less theoretical and applied interest than are mean-square and weak approximation; therefore we will not consider them in this book.

The works [4], [11] have used numerical integration of stochastic differential equations in the sense of weak approximation of solutions for approximately computing Wiener integrals by the Monte-Carlo method.

We will now give a short account of each Chapter in this book.

In the first Chapter we study numerical integration methods for stochastic systems in the sense of mean-square approximation. A fundamental role in the foundations of all constructions is played by the theorem on the relation between approximation on a finite interval and one-step approximation. Using this theorem we can prove the mean-square convergence of some or other method and establish the order of this convergence (§1). The conditions of this theorem use both properties of *mean* as well as *mean-square deviation* from one-step approximation. The theorem asserts that if p_1 and p_2 are the orders of accuracy of the one-step method for, respectively, the deviation of mathematical expectation and the mean-square deviation, and if $p_2 > 1/2$, $p_1 \geq p_2 + 1/2$, then the method converges and the order of approximation is $p_2 - 1/2$. In §5 we prove a strengthened convergence theorem, in which instead of (0.22) we obtain

$$\left(\mathbf{E}\max_k |X(t_k) - X_k|^2\right)^{1/2} \le C\, h^p. \tag{0.29}$$

In the deterministic theory, expansion of the solution by Taylor's formula underlies all one-step methods, both implicit and explicit. The analog of this expansion for stochastic systems–*Wagner–Platen expansion*–is considered in §2. Here, on the basis of the same convergence theorem, we establish the order of the method in dependence of the components in the expansion. Special attention is given to the construction of implicit methods. In the next Section (§3) we consider systems with additive noises, for which we construct several methods of order 3/2: explicit, implicit, Runge–Kutta type, and difference methods. Many results in this Section can, in principle, be transferred to general systems and be used for constructing methods of higher order of accuracy. We have to admit that the complexity of the methods increases sharply. On the basis of the relation between numerical integration problems and filtering problems with discrete information arrival, we construct in §4 optimal numerical integration methods for linear systems with additive noises.

In the second Chapter we consider means for a precise as well as approximate modeling of Itô integrals depending on one (§6) or several (§7) noises. It will turn out that the possibilities for exact modeling are, especially in the case of several noises, very limited. Therefore our main interest is in methods of approximate modeling. We will develop such methods in great detail for those Itô integrals that are necessary in the construction of *mean-square approximations* with first order of accuracy. In turn, the methods given lean on approximate numerical integration methods for concrete linear stochastic systems that are in a certain manner related to the Itô integrals being modeled.

Mean-square approximation is very laborious, in particular in view of the necessity of modeling Itô integrals which, as follows from Chapter 2, is a very complicated problem.

In the third Chapter we construct various methods of second order of accuracy in the weak sense for general systems of stochastic differential equations, as well as methods of third order of accuracy for *systems with additive noises*. These methods use random variables that are simple to model. They are simpler than mean-square ones, and we need compute substantially fewer operators in the coefficients of the system to obtain the same weak order of accuracy.

The fact that *weak approximations* of the solution suffice for the equations of mathematical physics (in [54] such approximations are called Monte-Carlo approximations) shows that precisely such approximations are of most interest in applications and should thus be in the center when investigating stochastic differential equations by numerical integration. Moreover, it must be stressed that in the construction of weak approximations we use mean-square approximations in an essential way.

In §8 we give a detailed construction of a one-step approximation of third order of accuracy. It is basic in constructing methods of second order of accuracy for stochastic systems of general type. In §9 we prove a theorem stating that if a one-step approximation has $(p+1)$th order of accuracy, then approximation on a finite interval has pth order of accuracy. This theorem plays the same role in the theory of weak approximation as the main convergence theorem does in the theory of mean-square approximation. In §10 we obtain a method of third order of accuracy for systems with additive noises, while in §11 we construct implicit methods for weak approximation. We have to note that up till now we have not found a way of constructing efficient Runge–Kutta type methods in the case of weak approximations. The examples of such methods in the existing literature have not completely escaped from computing derivatives of the coefficients of the systems, and therefore the problem of constructing Runge–Kutta methods is an actual one. Not less pressing is the construction of sufficiently general and efficient multi-step methods for both mean-square and weak approximation. Finally, the last Section (§12) of Chapter 3 is devoted to an important question related with the error of the Monte-Carlo method. This error can be estimated by the quantity $[\mathbf{V}f(\overline{X}(t_0+T))/N]^{1/2}$, which is close to $[\mathbf{V}f(X(t_0+T))/N]^{1/2}$. If the variance $\mathbf{V}f(X(t_0+T))$ is large, then we have to count a large number of trajectories. In §12 we introduce a system of stochastic equations with control, with solution including on a component Z, and being such that always, independent of the choice of the control,

$$\mathbf{E}Z(t_0+T) = \mathbf{E}f(X(t_0+T)), \tag{0.30}$$

while already $\mathbf{V}Z$ depends on the control. Since our aim is to compute $\mathbf{E}f(X(t_0+T))$, by (0.30) we can replace the modeling of $X(t_0+T)$ by that of Z such that for a suitable control $\mathbf{V}Z$ is reduced, and with it the error in the Monte-Carlo method. We can prove that under natural assumptions the optimal control makes $\mathbf{V}Z$ vanish. Although it is not guaranteed that we can find the optimal control (the problem of finding it is far more complicated than the initial problem of computing $\mathbf{E}f(X(t_0+T))$), we have nevertheless established the principal possibility of reducing the error of the Monte-Carlo method by choosing a suitable control.

On the basis of the relation between certain important classes of *Wiener integrals* and stochastic systems of equations we develop new approximate methods for computing Wiener integrals in the fourth Chapter. The corresponding systems of equations have a special form, and we construct methods of weak approximation for them. In §13 we consider Wiener integrals of functionals of integral type, and find methods of second order of accuracy for them. In §14 we consider integrals of a more particular, but often encountered, form (integrals of functionals of exponential type). For these we succeed in constructing efficient methods of already fourth order of accuracy. In this, final, Section our main attention goes to reducing the probabilistic error of the Monte-Carlo method (in the spirit of §12) and we finally give numerical results that fit in nicely with the theoretical results.

The references given in this book relate to the time at which the book was published in Russian, since the translation is identical with the original. An extensive list of references can be found in P.E. Kloeden and E. Platen, *Numerical Solution of Stochastic Differential Equations*, Springer, 1992.

The notation used in this book is the one generally accepted; it is explained in the course of presentation. Here we just note that the various constants in the text have been given the same letter K. In connection with this, instead of, e.g., K^2, $K + K$, $K - K$, $2K$, etc., we have simply written K.

CHAPTER 1

Mean-square approximation of solutions of systems of stochastic differential equations

1. Theorem on the order of convergence (theorem on the relation between approximation on a finite interval and one-step approximation)

1.1. Statement of the theorem. Let $(\Omega, \mathcal{F}, \mathbf{P})$ be a probability space, let $\mathcal{F}_t$, $t_0 \le t \le t_0 + T$, be a nonincreasing family of σ-subalgebras of $\mathcal{F}$, and let $(w_r(t), \mathcal{F}_t)$, $r = 1, \dots, q$, be independent Wiener processes. Consider the *system of stochastic differential equations in the sense of Itô*

$$dX = a(t, X)\,dt + \sum_{r=1}^{q} \sigma_r(t, X)\,dw_r(t), \tag{1.1}$$

where X, a, σ_r are vectors of dimension n.

Assume that the functions $a(t,x)$ and $\sigma_r(t,x)$ are defined and continuous for $t \in [t_0, t_0 + T]$, $x \in \mathbb{R}^n$ and satisfy a Lipschitz condition: for all $t \in [t_0, t_0 + T]$, $x \in \mathbb{R}^n$, $y \in \mathbb{R}^n$ there is an inequality

$$|a(t,x) - a(t,y)| + \sum_{r=1}^{q} |\sigma_r(t,x) - \sigma_r(t,y)| \le K\,|x - y|\,. \tag{1.2}$$

Here and below $|x|$ denotes the Euclidean norm of the vector x, and we denote by xy the scalar (inner) product of two vectors x and y.

Let $(X(t), \mathcal{F}(t))$, $t_0 \le t \le t_0 + T$, be a solution of the system (1.1) with $\mathbf{E}\,|X(t_0)|^2 < \infty$. The one-step approximation $\overline{X}_{t,x}(t+h)$, $t_0 \le t < t + h \le t_0 + T$, is defined as follows, and depends on x, t, h, and $\{w_1(\theta) - w_1(t), \dots, w_q(\theta) - w_q(t)\colon t \le \theta \le t + h\}$:

$$\overline{X}_{t,x}(t+h) = x + A(t, x, h; w_i(\theta) - w_i(t), i = 1, \dots, q, t \le \theta \le t + h). \tag{1.3}$$

Using the one-step approximation we recurrently construct the approximations $(\overline{X}_k, \mathcal{F}_{t_k})$, $k = 0, \dots, N$, $t_{k+1} - t_k = h_{k+1}$, $t_N = t_0 + T$:

$$\overline{X}_0 = X_0 = X(t_0),$$

$$\begin{aligned} \overline{X}_{k+1} &= \overline{X}_{t_k, \overline{X}_k}(t_{k+1}) \\ &= \overline{X}_k + A(t_k, \overline{X}_k, h_{k+1}; w_i(\theta) - w_i(t_k), i = 1, \dots, q, t_k \le \theta \le t_{k+1}). \end{aligned} \tag{1.4}$$

We will use the following notations. As in the Introduction, an approximation to $X(t_k)$ will be denoted by $\overline{X}(t_k)$, $\overline{X}_k$, or simply by X_k. Everywhere below we have put $\overline{X}_0 = X(t_0)$. Further, let X be an $\mathcal{F}_{t_k}$-measurable random variable with $\mathbf{E}|X|^2 < \infty$; as usual, $X_{t_k,X}(t)$ denotes the solution of the system (1.1) for $t_k \le t \le t_0+T$ satisfying the following initial condition at $t = t_k$: $X(t_k) = X$. By $\overline{X}_{t_k,X}(t_i)$, $t_i \ge t_k$, we denote an approximation of the solution at step i and such that $\overline{X}_k = X$. For example,

$$\overline{X}_{k+1} = \overline{X}_{t_k,\overline{X}_k}(t_{k+1}) = \overline{X}_{t_0,X_0}(t_{k+1}).$$

For simplicity reasons we assume that $t_{k+1} - t_k = h = T/N$.

THEOREM 1.1. *Suppose the one-step approximation $\overline{X}_{t,x}(t+h)$ has order of accuracy p_1 for the mathematical expectation of the deviation and order of accuracy p_2 for the mean-square deviation; more precisely, for arbitrary $t_0 \le t \le t_0 + T - h$, $x \in \mathbb{R}^n$ the following inequalities hold:*

$$\left|\mathbf{E}(X_{t,x}(t+h) - \overline{X}_{t,x}(t+h))\right| \le K(1+|x|^2)^{1/2}h^{p_1}, \tag{1.5}$$

$$\left[\mathbf{E}\left|X_{t,x}(t+h) - \overline{X}_{t,x}(t+h)\right|^2\right]^{1/2} \le K(1+|x|^2)^{1/2}h^{p_2}. \tag{1.6}$$

Also, let

$$p_2 \ge \frac{1}{2}, \qquad p_1 \ge p_2 + \frac{1}{2}. \tag{1.7}$$

Then for any N and $k = 0, 1, \ldots, N$ the following inequality holds:

$$\left[\mathbf{E}\left|X_{t_0,X_0}(t_k) - \overline{X}_{t_0,X_0}(t_k)\right|^2\right]^{1/2} \le K(1+\mathbf{E}|X_0|^2)^{1/2}h^{p_2-1/2}, \tag{1.8}$$

i.e. the order of accuracy of the method constructed using the one-step approximation $\overline{X}_{t,x}(t+h)$ is $p = p_2 - 1/2$.

We note that all constants K mentioned above, as well as the ones that will appear in the sequel, depend in the final analysis only on the system (1.1) and the approximations (1.3) and do not depend on X_0 and N.

1.2. Lemmas.

LEMMA 1.1. *There is a representation*

$$X_{t,x}(t+h) - X_{t,y}(t+h) = x - y + Z \tag{1.9}$$

for which

$$\mathbf{E}\left|X_{t,x}(t+h) - X_{t,y}(t+h)\right|^2 \le |x-y|^2(1+Kh), \tag{1.10}$$

$$\mathbf{E}Z^2 \le K\left|x-y\right|^2 h. \tag{1.11}$$

PROOF. Itô's formula readily implies that for $0 \leq \theta \leq h$:

$$\mathbf{E}\,|X_{t,x}(t+\theta) - X_{t,y}(t+\theta)|^2 = |x-y|^2$$
$$+ 2\mathbf{E}\int_t^{t+\theta} (X_{t,x}(s) - X_{t,y}(s))\,(a(s,X_{t,x}(s)) - a(s,X_{t,y}(s)))\,ds$$
$$+ \mathbf{E}\int_t^{t+\theta}\sum_{r=1}^{q} |\sigma_r(s,X_{t,x}(s)) - \sigma_r(s,X_{t,y}(s))|^2\,ds.$$

By (1.2) this implies

$$\mathbf{E}\,|X_{t,x}(t+\theta) - X_{t,y}(t+\theta)|^2 \leq |x-y|^2 + K\int_t^{t+\theta} \mathbf{E}\,|X_{t,x}(s) - X_{t,y}(s)|^2\,ds. \quad (1.12)$$

In turn, this implies

$$\mathbf{E}\,|X_{t,x}(t+\theta) - X_{t,y}(t+\theta)|^2 \leq |x-y|^2 \cdot e^{Kh}, \qquad 0 \leq \theta \leq h, \quad (1.13)$$

from which (1.10) follows. Further, since

$$Z = \int_t^{t+h}\sum_{r=1}^{q} (\sigma_r(s,X_{t,x}(s)) - \sigma_r(s,X_{t,y}(s)))\,dw_r(s)$$
$$+ \int_t^{t+h} (a(s,X_{t,x}(s)) - a(s,X_{t,y}(s)))\,ds,$$

it is not difficult to obtain (1.11), using (1.13). □

REMARK 1.1. In the sequel we will use a *conditional version of the inequalities* (1.5), (1.6), (1.10), (1.11). In this conditional version the deterministic variables x, y are replaced by $\mathcal{F}_t$-measurable random variables X, Y. For example, the conditional version of (1.5) reads:

$$\left|\mathbf{E}\left(X_{t,X}(t+h) - \overline{X}_{t,X}(t+h) \mid \mathcal{F}_t\right)\right| \leq K(1+|X|^2)^{1/2}h^{p_1}. \quad (1.14)$$

We will also use simple consequences of these inequalities. For example, (1.14) implies

$$\mathbf{E}\left|\mathbf{E}\left(X_{t,X}(t+h) - \overline{X}_{t,X}(t+h) \mid \mathcal{F}_t\right)\right|^2 \leq K(1+\mathbf{E}|X|^2)^{1/2}h^{2p_1}. \quad (1.15)$$

The proof of these conditional versions rests on an assumption of the following kind: if ζ is $\tilde{\mathcal{F}}$-measurable, $\tilde{\mathcal{F}} \subset \mathcal{F}$, $f(x,\omega)$ does not depend on $\tilde{\mathcal{F}}$, $\omega \in \Omega$, and $\mathbf{E}f(x,\omega) = \phi(x)$, then $\mathbf{E}(f(\zeta,\omega) \mid \tilde{\mathcal{F}}) = \phi(\zeta)$ (see [8, p. 67], [21, p. 158]). In the case under consideration the *increments of the Wiener processes* $w_1(\theta) - w_1(t), \dots, w_q(\theta) - w_q(t)$, $t \leq \theta \leq t+h$, do not depend on $\mathcal{F}_t$. So, $X_{t,x}(t+h)$ does not depend on $\mathcal{F}_t$ (see, e.g., [8, p. 67]), and neither does $\overline{X}_{t,x}(t+h)$, which is formed such that it depends only on x, t, h, and the increments above.

LEMMA 1.2. *For all natural numbers N and all $k = 0, \ldots, N$ the following inequality holds:*

$$\mathbf{E}\left|\overline{X}_k\right|^2 \le K\left(1 + \mathbf{E}|X_0|^2\right). \tag{1.16}$$

PROOF. Suppose that $\mathbf{E}\left|\overline{X}_k\right|^2 < \infty$. Then, using the conditional version of (1.16), we obtain

$$\mathbf{E}\left|X_{t_k,\bar{X}_k}(t_{k+1}) - \overline{X}_{t_k,\bar{X}_k}(t_{k+1})\right|^2 \le K^2\left(1 + \mathbf{E}\left|\overline{X}_k\right|^2\right)^2 h^{2p_2}. \tag{1.17}$$

It is well known (see [8, p. 48]) that if an $\mathcal{F}_t$-measurable random variable X has bounded second moment, then the solution $X_{t,X}(t+\theta)$ also has bounded second moment. Therefore $\mathbf{E}\left|X_{t_k,\bar{X}_k}(t_{k+1})\right| < \infty$. This and (1.17) readily imply that $\mathbf{E}\left|\overline{X}_{k+1}\right|^2 < \infty$ (recall that $\overline{X}_{t_k,\bar{X}_k}(t_{k+1}) = \overline{X}_{k+1}$). Since $\mathbf{E}\left|\overline{X}_0\right|^2 < \infty$, we have hence proved the existence of all $\mathbf{E}\left|\overline{X}_k\right|^2 < \infty$, $k = 0, \ldots, N$.

Consider the equation

$$\begin{aligned}\mathbf{E}\left|\overline{X}_{k+1}\right|^2 &= \mathbf{E}\left|\overline{X}_k\right|^2 + \mathbf{E}\left|X_{t_k,\bar{X}_k}(t_{k+1}) - \overline{X}_k\right|^2 + \mathbf{E}\left|X_{t_k,\bar{X}_k}(t_{k+1}) - \overline{X}_{t_k,\bar{X}_k}(t_{k+1})\right|^2 \\ &\quad + 2\mathbf{E}\overline{X}_k\left(X_{t_k,\bar{X}_k}(t_{k+1}) - \overline{X}_k\right) + 2\mathbf{E}\overline{X}_k\left(\overline{X}_{t_k,\bar{X}_k}(t_{k+1}) - X_{t_k,\bar{X}_k}(t_{k+1})\right) \\ &\quad + 2\mathbf{E}\left(X_{t_k,\bar{X}_k}(t_{k+1}) - \overline{X}_k\right)\left(\overline{X}_{t_k,\bar{X}_k}(t_{k+1}) - X_{t_k,\bar{X}_k}(t_{k+1})\right).\end{aligned} \tag{1.18}$$

We have (see [8, p. 48])

$$\mathbf{E}\left|X_{t_k,\bar{X}_k}(t_{k+1}) - \overline{X}_k\right|^2 \le K\left(1 + \mathbf{E}\left|\overline{X}_k\right|^2\right) h. \tag{1.19}$$

Further, from (1.17) and (1.19) we obtain

$$\begin{aligned}&2\left|\mathbf{E}\left(X_{t_k,\bar{X}_k}(t_{k+1}) - \overline{X}_k\right)\left(X_{t_k,\bar{X}_k}(t_{k+1}) - \overline{X}_{t_k,\bar{X}_k}(t_{k+1})\right)\right| \\ &\le 2\left[\mathbf{E}\left|X_{t_k,\bar{X}_k}(t_{k+1}) - \overline{X}_k\right|^2\right]^{1/2}\left[\mathbf{E}\left|X_{t_k,\bar{X}_k}(t_{k+1}) - \overline{X}_{t_k,\bar{X}_k}(t_{k+1})\right|^2\right]^{1/2} \\ &\le K\left(1 + \mathbf{E}\left|\overline{X}_k\right|^2\right) h^{p_2+1/2}.\end{aligned} \tag{1.20}$$

It is not difficult to prove the inequality

$$\mathbf{E}\left|\mathbf{E}\left(X_{t_k,\bar{X}_k}(t_{k+1}) - \overline{X}_k \mid \mathcal{F}_{t_k}\right)\right|^2 \le K\left(1 + \mathbf{E}\left|\overline{X}_k\right|^2\right) h^2. \tag{1.21}$$

Therefore

$$\begin{aligned}2\left|\mathbf{E}\overline{X}_k\left(X_{t_k,\bar{X}_k}(t_{k+1}) - \overline{X}_k\right)\right| &= 2\left|\mathbf{E}\overline{X}_k\mathbf{E}\left(X_{t_k,\bar{X}_k}(t_{k+1}) - \overline{X}_k \mid \mathcal{F}_{t_k}\right)\right| \\ &\le 2\left(\mathbf{E}\left|\overline{X}_k\right|^2\right)^{1/2}\left(\mathbf{E}\left|\mathbf{E}\left(X_{t_k,\bar{X}_k}(t_{k+1}) - \overline{X}_k \mid \mathcal{F}_{t_k}\right)\right|^2\right)^{1/2} \\ &\le K\left(1 + \mathbf{E}\left|\overline{X}_k\right|^2\right) h.\end{aligned} \tag{1.22}$$

Similarly, but referring to (1.15) instead of (1.21), we obtain

$$2\left|\mathbf{E}\overline{X}_k\left(X_{t_k,\bar{X}_k}(t_{k+1})-\overline{X}_{t_k,\bar{X}_k}(t_{k+1})\right)\right|\leq K\left(1+\mathbf{E}\left|\overline{X}_k\right|^2\right)h^{p_1}. \tag{1.23}$$

Applying the inequalities (1.19), (1.17) (1.22), (1.23), and (1.20) to the equality (1.18) and recalling that $p_1 \geq 1$, $p_2 \geq 1/2$, we arrive at the inequality (taking, without loss of generality, $h \leq 1$):

$$\mathbf{E}\left|\overline{X}_{k+1}\right|^2\leq\mathbf{E}\left|\overline{X}_k\right|^2+K\left(1+\mathbf{E}\left|\overline{X}_k\right|^2\right)h=(1+Kh)\mathbf{E}\left|\overline{X}_k\right|^2+Kh. \tag{1.24}$$

Hence, using a well-known result which, for the sake of reference, is stated as Lemma 1.3 below, we obtain (1.16). □

LEMMA 1.3. *Suppose that for arbitrary N and $k = 0,\dots,N$ we have*

$$u_{k+1}\leq(1+Ah)u_k+Bh^p, \tag{1.25}$$

where $h = T/N$, $A \geq 0$, $B \geq 0$, $p \geq 1$, $u_k \geq 0$, $k = 0,\dots,N$. Then

$$u_k\leq e^{AT}u_0+\frac{B}{A}\left(e^{AT}-1\right)h^{p-1} \tag{1.26}$$

(where for $A = 0$ we put $\left(e^{AT}-1\right)/A$ equal to zero).

1.3. Proof of Theorem 1.1.

$$\begin{aligned}X_{t_0,X_0}(t_{k+1})-\overline{X}_{t_0,X_0}(t_{k+1})&=X_{t_k,X(t_k)}(t_{k+1})-\overline{X}_{t_k,\bar{X}_k}(t_{k+1})\\&=\left(X_{t_k,X(t_k)}(t_{k+1})-X_{t_k,\bar{X}_k}(t_{k+1})\right)\\&\quad+\left(X_{t_k,\bar{X}_k}(t_{k+1})-\overline{X}_{t_k,\bar{X}_k}(t_{k+1})\right).\end{aligned} \tag{1.27}$$

The first difference at the righthand side of (1.27) is the error of the solution arising because of the error in the initial data at time t_k, accumulated at the kth step. The second difference is the one-step error at the $(k+1)$th step. Taking the square of both sides of the equation, we obtain

$$\begin{aligned}\mathbf{E}\left|X_{t_0,X_0}(t_{k+1})-\overline{X}_{t_0,X_0}(t_{k+1})\right|^2&=\mathbf{E}\mathbf{E}\left(\left(X_{t_k,X(t_k)}(t_{k+1})-X_{t_k,\bar{X}_k}(t_{k+1})\right)^2\mid\mathcal{F}_{t_k}\right)\\&\quad+\mathbf{E}\mathbf{E}\left(\left(X_{t_k,\bar{X}_k}(t_{k+1})-\overline{X}_{t_k,\bar{X}_k}(t_{k+1})\right)^2\mid\mathcal{F}_{t_k}\right)\\&\quad+2\mathbf{E}\mathbf{E}\left(\left(X_{t_k,X(t_k)}(t_{k+1})-X_{t_k,\bar{X}_k}(t_{k+1})\right)\right.\\&\quad\left.\times\left(X_{t_k,\bar{X}_k}(t_{k+1})-\overline{X}_{t_k,\bar{X}_k}(t_{k+1})\right)\mid\mathcal{F}_{t_k}\right).\end{aligned} \tag{1.28}$$

By the conditional version of Lemma 1.1 we have

$$\mathbf{E}\mathbf{E}\left(\left(X_{t_k,X(t_k)}(t_{k+1})-X_{t_k,\bar{X}_k}(t_{k+1})\right)^2\mid\mathcal{F}_{t_k}\right)\leq\mathbf{E}\left|X(t_k)-\overline{X}_k\right|^2\cdot(1+Kh). \tag{1.29}$$

By the conditional version of (1.6) and Lemma 1.2 we have

$$\begin{aligned}\mathbf{E}\mathbf{E}\left(\left(X_{t_k,\bar{X}_k}(t_{k+1})-\overline{X}_{t_k,\bar{X}_k}(t_{k+1})\right)^2 \mid \mathcal{F}_{t_k}\right) &\le K\left(1+\mathbf{E}\left|\overline{X}_k\right|^2\right)h^{2p_2}\\ &\le K\left(1+\mathbf{E}\left|X_0\right|^2\right)h^{2p_2}.\end{aligned} \tag{1.30}$$

The difference $X_{t_k,X(t_k)}(t_{k+1}) - X_{t_k,\bar{X}_k}(t_{k+1})$ in the last summand in (1.28) can be treated using Lemma 1.1. We then obtain two terms each of which can be estimated individually. Using (1.15) and Lemma 1.2 we obtain

$$\begin{aligned}&\left|\mathbf{E}\mathbf{E}\left(\left(X(t_k)-\overline{X}_k\right)\left(X_{t_k,\bar{X}_k}(t_{k+1})-\overline{X}_{t_k,\bar{X}_k}(t_{k+1})\right) \mid \mathcal{F}_{t_k}\right)\right|\\ &\quad=\left|\mathbf{E}\left(\left(X(t_k)-\overline{X}_k\right)\mathbf{E}\left(X_{t_k,\bar{X}_k}(t_{k+1})-\overline{X}_{t_k,\bar{X}_k}(t_{k+1}) \mid \mathcal{F}_{t_k}\right)\right)\right|\\ &\quad\le\left(\mathbf{E}\left|X(t_k)-\overline{X}_k\right|^2\right)^{1/2} K\left(1+\mathbf{E}\left|X_0\right|^2\right)^{1/2} h^{p_1}.\end{aligned} \tag{1.31}$$

Finally, using Lemma 1.2 and (1.6) we obtain

$$\begin{aligned}&\left|\mathbf{E}\left(Z\left(X_{t_k,\bar{X}_k}(t_{k+1})-\overline{X}_{t_k,\bar{X}_k}(t_{k+1})\right)\right)\right|\\ &\quad\le\left[\mathbf{E}\mathbf{E}\left(Z^2 \mid \mathcal{F}_{t_k}\right)\right]^{1/2}\left[\mathbf{E}\mathbf{E}\left(\left(X_{t_k,\bar{X}_k}(t_{k+1})-\overline{X}_{t_k,\bar{X}_k}(t_{k+1})\right)^2 \mid \mathcal{F}_{t_k}\right)\right]^{1/2}\\ &\quad\le K\left(\mathbf{E}\left|X(t_k)-\overline{X}_k\right|^2\right)^{1/2}\left(1+\mathbf{E}\left|X_0\right|^2\right)^{1/2} h^{p_2+1/2}.\end{aligned} \tag{1.32}$$

We introduce the notation $\epsilon_k^2 = \mathbf{E}\left|X(t_k)-\overline{X}_k\right|^2$. Relations (1.28)–(1.32) and the condition $p_1 \ge p_2 + 1/2$ then lead to the inequality ($h < 1$)

$$\epsilon_{k+1}^2 \le \epsilon_k^2(1+Kh) + K\left(1+\mathbf{E}\left|X_0\right|^2\right)^{1/2}\epsilon_k h^{p_2+1/2} + K\left(1+\mathbf{E}\left|X_0\right|^2\right)h^{2p_2}.$$

Using the elementary relation

$$\left(1+\mathbf{E}\left|X_0\right|^2\right)^{1/2}\epsilon_k h^{p_2+1/2} \le \frac{\epsilon_k^2 k}{2} + \frac{1+\mathbf{E}\left|X_0\right|^2}{2}h^{2p_2}$$

we obtain

$$\epsilon_{k+1}^2 \le \epsilon_k^2(1+Kh) + K\left(1+\mathbf{E}\left|X_0\right|^2\right)h^{2p_2}.$$

Inequality (1.8) follows from this, taking into account Lemma 1.3 and the fact that $\epsilon_0^2 = 0$. This proves Theorem 1.1.

1.4. Discussion. The rule:

> if, in a single step, the mean-square error has order h^{p_2} (i.e. inequality (1.6) holds), then it has order $h^{p_2-1/2}$ on the whole interval

is not true without the additional condition $p_1 \ge p_2 + 1/2$. A simple example in which this can be seen is given by the method

$$\overline{X}_{k+1} = \overline{X}_k + \sigma(t_k, \overline{X}_k)\Delta_k w(h)$$

for the scalar version of the system (1.1). It is easy to see that here $p_2 = 1$ while the method diverges for $a \neq 0$.

The rule:

> if, in a single step, the mean-square deviation has order h^{p_2}, then it has order h^{p_2-1} on the whole interval

is true, but a bit rough. Following this rule we cannot prove the convergence of the Euler method, in which $p_2 = 1$. Moreover, a more efficient rule for the mean-square deviation cannot be found if we are guided only by the mean-square characteristic of the one-step approximation. The rule following from Theorem 1.1 is based on properties of both the *mean* as well as the *mean-square deviation* of the one-step approximation. In particular, for Euler's method $p_1 = 2$, $p_2 = 1$, and so it follows from Theorem 1.1 that Euler's method has order of accuracy 1/2.

Properties of the mean are used at one place in the proof of the Theorem (and at a 'delicate' place indeed); more precisely, when deriving the inequality (1.31). If the lefthand side of this inequality is roughly estimated, without taking into account (1.5), using the Bunyakovsky–Schwarz inequality we obtain

$$\begin{aligned}
&\left|\mathbf{E}\mathbf{E}\left(\left(X(t_k)-\overline{X}_k\right)\left(X_{t_k,\bar{X}_k}(t_{k+1})-\overline{X}_{t_k,\bar{X}_k}(t_{k+1})\right) \mid \mathcal{F}_{t_k}\right)\right| \\
&\quad \leq \left[\mathbf{E}\left(X(t_k)-\overline{X}_k\right)^2\right]^{1/2}\left[\mathbf{E}\left(X_{t_k,\bar{X}_k}(t_{k+1})-\overline{X}_{t_k,\bar{X}_k}(t_{k+1})\right)^2\right]^{1/2} \\
&\quad \leq \left[\mathbf{E}\left(X(t_k)-\overline{X}_k\right)^2\right]^{1/2} K\left(1+\mathbf{E}\left|\overline{X}_k\right|^2\right)^{1/2} h^{p_2}.
\end{aligned}$$

While at the righthand side of (1.31) we had the factor h^{p_1}, here we only have h^{p_2}, which is not sufficient for concluding the Theorem.

1.5. Equations in the sense of Stratonovich. Consider the following system of equations in the sense of Stratonovich:

$$dX = a(t,X)\,dt + \sum_{r=1}^{q} \sigma_r(t,X) \star dw_r(t), \tag{1.33}$$

where, in distinction to (1.1), we use the sign $\star$. It is well known (see, e.g., [41, p. 219]) that this system is equivalent to the following system in the sense of Itô:

$$dX = \left(a(t,X) + \frac{1}{2}\sum_{r=1}^{q} \frac{\partial \sigma_r}{\partial x}(t,X)\sigma_r(t,X)\right) dt + \sum_{r=1}^{q} \sigma_r(t,X)\,dw_r(t). \tag{1.34}$$

In this system, $\partial\sigma_r/\partial x$ is the matrix with entry $\partial\sigma_r^i/\partial x^j$ at the intersection of the ith row and jth column. Here, of course, we assume the σ_r, $r = 1,\dots,q$, to be not only differentiable, but also require that the vectors $(\partial\sigma_r/\partial x)\sigma_r$ satisfy a uniform Lipschitz condition with respect to $x \in \mathbb{R}^n$, i.e. that the system (1.34) satisfies a condition (1.2).

By the above, it is not difficult to understand that Theorem 1.1 remains true for solutions of equations understood in the sense of Stratonovich.

1.6. Euler's method. For the system (1.1) we consider the one-step approximation (1.3) of the form

$$\overline{X}_{t,x}(t+h) = x + a(t,x)h + \sum_{r=1}^{q} \sigma_r(t,X)\Delta_t w_r(h), \tag{1.35}$$

where $\Delta_t w_r(h) = w_r(t+h) - w_r(t)$.

By (1.4), this approximation generates the Euler method:

$$X_{k+1} = X_k + a_k h + \sum_{r=1}^{q} \sigma_{r_k} \Delta_k w_r(h), \tag{1.36}$$

where a_k, σ_{r_k} are the values of the coefficients a and σ_r at the point (t_k, X_k), and $\Delta_k w_r(h) = w_r(t_k+h) - w_r(t_k)$. For (1.35) we can find p_1 and p_2 satisfying the estimates (1.5) and (1.6). For this we termwise subtract (1.36) from the identity

$$X_{t,x}(t+h) = x + \int_t^{t+h} a(s, X_{t,x}(s))\, ds + \sum_{r=1}^{q} \sigma_r(s, X_{t,x}(s))\, dw_r(s). \tag{1.37}$$

We obtain

$$\begin{aligned} X_{t,x}(t+h) - \overline{X}_{t,x}(t+h) = &\int_t^{t+h} (a(s, X_{t,x}(s)) - a(t,x))\, ds \\ &+ \sum_{r=1}^{q} \int_t^{t+h} (\sigma_r(s, X_{t,x}(s)) - \sigma_r(t,x))\, dw_r(s). \end{aligned} \tag{1.38}$$

Taking mathematical expectations leads to

$$\mathbf{E}\left(X_{t,x}(t+h) - \overline{X}_{t,x}(t+h)\right) = \mathbf{E}\int_t^{t+h} (a(s, X_{t,x}(s)) - a(t,x))\, ds. \tag{1.39}$$

Taking the square of both sides of (1.38) (i.e. taking the scalar product of the vectors with themselves) and then taking mathematical expectations leads to

$$\mathbf{E}\left(X_{t,x}(t+h)-\overline{X}_{t,x}(t+h)\right)^2 = \mathbf{E}\left(\int_t^{t+h} (a(s,X_{t,x}(s))-a(t,x))\,ds\right)^2$$
$$+2\mathbf{E}\left(\int_t^{t+h} (a(s,X_{t,x}(s))-a(t,x))\,ds \sum_{r=1}^q \int_t^{t+h} (\sigma_r(s,X_{t,x}(s))-\sigma_r(t,x))\,dw_r(s)\right)$$
$$+\sum_{r=1}^q \int_t^{t+h} \mathbf{E}\,(\sigma_r(s,X_{t,x}(s))-\sigma_r(t,x))^2\,ds$$
$$\leq \mathbf{E}\left(\int_t^{t+h} (a(s,X_{t,x}(s))-a(t,x))\,ds\right)^2$$
$$+2\left(\mathbf{E}\int_t^{t+h} (a(s,X_{t,x}(s))-a(t,x))^2\right)^{1/2}$$
$$\times\left(\sum_{r=1}^q \int_t^{t+h} \mathbf{E}\,(\sigma_r(s,X_{t,x}(s))-\sigma_r(t,x))^2\,ds\right)^{1/2}$$
$$+\sum_{r=1}^q \int_t^{t+h} \mathbf{E}\,(\sigma_r(s,X_{t,x}(s))-\sigma_r(t,x))^2\,ds$$
$$\leq 2\mathbf{E}\left(\int_t^{t+h} (a(s,X_{t,x}(s))-a(t,x))\,ds\right)^2$$
$$+2\sum_{r=1}^q \int_t^{t+h} \mathbf{E}\,(\sigma_r(s,X_{t,x}(s))-\sigma_r(t,x))^2\,ds. \tag{1.40}$$

We now assume that, in addition to (1.2), the functions $a(t,x)$ and $\sigma_r(t,x)$ have partial derivatives with respect to t that grow at most as a linear function of x as $|x|\to\infty$. Then a and σ_r satisfy an inequality of the form

$$|a(s,X_{t,x}(s))-a(t,x)| \leq |a(s,X_{t,x}(s))-a(s,x)|+|a(s,x)-a(t,x)|$$
$$\leq K\,|X_{t,x}(s)-x|+K\left(1+|x|^2\right)^{1/2}(s-t). \tag{1.41}$$

By the Bunyakovsky–Schwarz inequality we have

$$\left|\int_t^{t+h} (a(s,X_{t,x}(s))-a(t,x))\,ds\right|^2 \leq h\int_t^{t+h} (a(s,X_{t,x}(s))-a(t,x))^2\,ds,$$

so that by (1.41) and (1.19) the first term at the righthand side of (1.40) is bounded by $K\,(1+|x|^2)\,h^3$. Using the inequality (1.41) for σ_r and (1.19), we see that the second term at the righthand side of (1.40) is bounded by $K(1+|x|^2)\,h$. Thus, $p_2 = 1$. Note that this value of p_2 is determined by the second term at the righthand side of (1.40). To find p_1 we turn to (1.39). We assume in addition that the derivatives $\partial a^i/\partial x^j$ and $\partial^2 a^i/\partial x^j \partial x^k$, $i,j,k = 1,\dots,n$, are uniformly bounded, and we write

$$a(s, X_{t,x}(s)) - a(t,x) = \frac{\partial a}{\partial x}(t,x)\,(X_{t,x}(s) - x) + a(s,x) - a(t,x) + \rho, \quad (1.42)$$

where

$$|\rho| \le K\,(X_{t,x}(s) - x)^2. \quad (1.43)$$

Since

$$\mathbf{E}\,(X_{t,x}(s) - x) = \mathbf{E}\int_t^s a(\theta, X_{t,x}(\theta))\,d\theta \quad (1.44)$$

and (1.2) for $y = 0$ implies that for all $t_0 \le \theta \le t_0 + T$ we have

$$|a(\theta, X_{t,x}(\theta))| \le |a(\theta,0)| + |a(\theta, X_{t,x}(\theta)) - a(\theta,0)| \le K + K\,|X_{t,x}(\theta)|,$$

it is not difficult to obtain the estimate

$$\left|\int_t^{t+h} \mathbf{E}\,(X_{t,x}(s) - x)\,ds\right| \le K\left(1 + |x|^2\right)^{1/2} h^2.$$

This and (1.39), (1.42), (1.43) imply (1.5) with $p_1 = 2$.

By Theorem 1.1, under the given assumptions regarding the coefficients a and σ_r, Euler's method is a method of order $p = p_2 - 1/2 = 1/2$. We now consider Euler's method for the following system with additive noises ($\sigma_r(t,x)$, $r = 1,\dots,q$, independent of x):

$$dX(t) = a(t,X)\,dt + \sum_{r=1}^{q} \sigma_r(t)\,dw_r(t). \quad (1.45)$$

It can be readily seen that in this case the second term at the righthand side of (1.40) can be bounded by $K(1+|x|^2)\,h^3$. As a result we find $p_2 = 3/2$. Since, as before, $p_1 = 2 \ge p_2 + 1/2$, by Theorem 1.1 the order of Euler's method for the system (1.45) is $p = p_2 - 1/2 = 1$.

REMARK 1.2. Formula (1.36) approximates the solution $X(t)$ of (1.1) at the nodes t_k. Consider a piecewise linear Euler approximation on the entire interval $t_0 \le t \le t_0 + T$:

$$\overline{X}(t) = X_k + a_k(t - t_k) + \sum_{r=1}^{q} \sigma_{r_k} \Delta_k w_r(h) \frac{t - t_k}{h}, \qquad t_k \le t \le t_{k+1}. \quad (1.46)$$

Note that now $\overline{X}(t)$ is not $\mathcal{F}_t$-measurable, since $w_r(t_{k+1})$, $r = 1,\dots,q$, participate in (1.46). Equation (1.46) immediately implies

$$\mathbf{E}\left(\overline{X}(t) - X_k\right)^2 = O(h), \qquad t_k \le t \le t_{k+1}.$$

Therefore, on the interval $t_0 \le t \le t_0 + T$ we have

$$\left(\mathbf{E}\left(\overline{X}(t) - X(t)\right)^2\right)^{1/2} = O(h^{1/2}),$$

i.e. a piecewise linear approximation has the same order of accuracy on the entire interval as in the nodes. It is necessary to stress that this fact only holds for methods whose order of accuracy does not exceed 1/2. In fact, consider the scalar equation

$$dX = dw, \qquad X(0) = 0.$$

For it the Euler approximation

$$X_0 = 0, \qquad X_{k+1} = X_k + \Delta_k w(h)$$

is exact at nodes, i.e. it has infinite order of accuracy at nodes. At the same time a piecewise linear approximation, e.g. for $t = (t_k + t_{k+1})/2$, has overall error of order 1/2:

$$\mathbf{E}\left(X\left(\frac{t_k + t_{k+1}}{2}\right) - \overline{X}\left(\frac{t_k + t_{k+1}}{2}\right)\right)^2 = \mathbf{E}\left(w\left(\frac{t_k + t_{k+1}}{2}\right) - \frac{1}{2}\left(w(t_{k+1}) + w(t_k)\right)\right)^2$$
$$= \frac{1}{4}h.$$

1.7. Examples.

EXAMPLE 1.1. Take a piecewise linear interpolation of the Wiener process, $w^h(t) = w(t_k) + \Delta w(h)(t - t_k)/h$, $t_k \le t \le t_{k+1}$, and consider, instead of the equation

$$dX = a(t, X)\,dt + \sigma(t, X)\,dw(t),$$

on each interval $t_k \le t \le t_{k+1}$ the equation

$$dX^h = a(t, X^h)\,dt + \sigma(t, X^h)\,dw^h(t)$$

and its solution at the nodes $X_k = X^h(t_k)$, which is $\mathcal{F}_{t_k}$-measurable. For the equation

$$dX = aX\,dt + \sigma X\,dw, \qquad 0 \le t \le T, \tag{1.47}$$

we obtain as one-step approximation:

$$\overline{X}_{t,x}(t+h) = e^{(a + \sigma(w(t+h) - w(t))/h)h} x. \tag{1.48}$$

At the same time,

$$X_{t,x}(t+h) = e^{(a - \sigma^2/2)h + \sigma(w(t+h) - w(t))} x. \tag{1.49}$$

Since

$$\mathbf{E}e^{\sigma(w(t+h) - w(t))} = e^{(\sigma^2/2)h},$$

we have

$$\mathbf{E}\left(X_{t,x}(t+h) - \overline{X}_{t,x}(t+h)\right) = \left(e^{ah} - e^{(a+\sigma^2/2)h}\right)x = O(h),$$

i.e. $p_1 = 1$ and Theorem 1.1 cannot be used to ensure convergence of the approximations. Actually, there is no convergence, since

$$\mathbf{E}X_{k+1} = \mathbf{E}e^{(a+\sigma(w(t_{k+1})-w(t_k))/h)h}\mathbf{E}X_k = e^{(a+\sigma^2/2)h}\mathbf{E}X_k,$$

$$\mathbf{E}\overline{X}(T) = \mathbf{E}X_N = e^{(a+\sigma^2/2)T}\mathbf{E}X_0,$$

while

$$\mathbf{E}X(T) = e^{aT}\mathbf{E}X_0.$$

Note that if, instead of (1.47), we consider an equation in the sense of Stratonovich,

$$dX = aX\,dt + \sigma X \star dw, \tag{1.50}$$

then the approximation (1.48) coincides with the solution (see the relation between (1.33) and (1.34)):

$$X_{t,x}(t+h) = e^{(a+\sigma(w(t+h)-w(t))/h)h}x,$$

i.e. the approximation (1.48) for (1.50) has infinite order of accuracy.

EXAMPLE 1.2. For equation (1.47) we consider the implicit method (trapezium method):

$$X_{k+1} = X_k + a\frac{X_k + X_{k+1}}{2}h + \sigma\frac{X_k + X_{k+1}}{2}\Delta_k w(h). \tag{1.51}$$

So,

$$X_{k+1} - X_k = \frac{aX_kh + \sigma X_k\Delta_k w(h)}{1 - \frac{ah}{2} - \frac{\sigma}{2}\Delta_k w(h)}. \tag{1.52}$$

Since the mathematical expectation of the righthand side does not exist, it is clear that the approximation (1.51) cannot converge in mean-square to the solution (1.47). Consider the following one-step approximation:

$$\overline{X}_{t,x}(t+h) = a + axh + \sigma x\Delta w(h) + \frac{\sigma^2}{2}x\Delta^2 w(h),$$
$$\Delta w(h) = w(t+h) - w(t), \tag{1.53}$$

which can be thought of as corresponding to (1.52). We have

$$\mathbf{E}\overline{X}_{t,x}(t+h) = \left(1 + ah + \frac{\sigma^2}{2}h\right)x,$$

which implies that $p_1 = 1$. It is not difficult to prove that also in this case there is no convergence of the approximation (1.53) to the solution of the equation (1.47). At the same time, an explicit computation gives that for the approximation (1.53) to the solution of (1.50) we have $p_1 = 2$, $p_2 = 3/2$, i.e. the method based on (1.53) has first order of accuracy for the equation (1.50). The computation can be performed by expanding the exponential $e^{ah+\sigma\Delta w(h)}$ in powers of h and $w(h)$ and retaining in all relations the powers up to h^2 and $(\Delta w(h))^4$ inclusive.

In particular, these examples warn against a noncritical use of 'sufficiently natural' methods.

2. Methods based on an analog of Taylor expansion of the solution

2.1. Taylor expansion of the solution for systems of ordinary differential equations. As is well known, the expansion just mentioned lies at the basis of all one-step methods, both implicit and explicit (e.g., Runge–Kutta type methods). Here we will give it in a form convenient for our subsequent construction of a stochastic analog.

Consider the system of ordinary differential equations

$$\frac{dX}{dt} = a(t, X). \tag{2.1}$$

The righthand side of (2.1) is assumed to be such that all subsequent constructions can be performed. For this it suffices that the function $a(t, x)$, $t_0 \le t \le t_0+T$, $x \in \mathbb{R}^n$, is sufficiently smooth and that $a(t, x)$ grows as an at most linear function of $|x|$ as $|x| \to \infty$.

Let $f(t, x)$ be a scalar or vector function (of course, sufficiently smooth). Along a solution $X(t)$ of (2.1) we have

$$\frac{d}{dt} f(t, X(t)) = \frac{\partial f}{\partial t}(t, X(t)) + \frac{\partial f}{\partial x}(t, X(t)) a(t, X(t)). \tag{2.2}$$

Let L be the operator (*Lyapunov operator*)

$$L = \frac{\partial}{\partial t} + a \frac{\partial}{\partial x}.$$

Assuming $X(t) = x$, (2.2) implies

$$f(s, X(s)) = f(t, x) + \int_t^s Lf(\theta, X(\theta))\, d\theta. \tag{2.3}$$

Let $f(t, x) = x$. Then $Lf(t, x) = a(t, x)$, $L^2 f(t, x) = La(t, x)$, etc. Therefore (2.3) for $s = t + h$ implies

$$X(t+h) = x + \int_t^{t+h} a(s, X(s))\, ds \tag{2.4}$$

(we have given these arguments for deriving the obvious identity (2.4) for uniformity with the subsequent computations). Further, using (2.3) for $a(s, X(s))$, we obtain

$$\begin{aligned} X(t+h) &= x + \int_t^{t+h} \left(a(t, x) + \int_t^s La(\theta, X(\theta))\, d\theta \right) ds \\ &= x + a(t, x)h + \int_t^{t+h} (t + h - s) La(s, X(s))\, ds. \end{aligned} \tag{2.5}$$

Again we use (2.3), but now for $La(s,X(s))$. We find

$$X(t+h) = x + a(t,x)h + La(t,x)\frac{h^2}{2} + \int_t^{t+h} \frac{(t+h-s)^2}{2} L^2 a(s,X(s))\,ds. \quad (2.6)$$

Continuing this way we obtain for the solution of (2.1) the well-known *Taylor expansion* in powers of h in a neighborhood of t. This expansion lies at the basis of creating explicit methods of various orders of accuracy, and reads:

$$\begin{aligned} X(t+h) &= x + a(t,x)h + La(t,x)\frac{h^2}{2} + \dots \\ &+ L^{m-1}a(t,x)\frac{h^m}{m!} + \int_t^{t+h} \frac{(t+h-s)^m}{m!} L^m a(s,X(s))\,ds. \end{aligned} \quad (2.7)$$

By (2.7), the *one-step approximation*

$$\overline{X}_{t,x}(t+h) = x + a(t,x)h + La(t,x)\frac{h^2}{2} + \dots + L^{m-1}a(t,x)\frac{h^m}{m!} \quad (2.8)$$

has error of order $m+1$ at a step, while the method based on (2.8) has mth order of accuracy.

We now give a similar method for obtaining a number of *implicit methods*. Recall that, in distinction to *explicit methods*, implicit methods have far better stability properties. They are necessary for integrating so-called stiff systems of differential equations (see, e.g., [39], [43]).

Using (2.3) we can write down the formula

$$f(s,X(s)) = f(t+h, X(t+h)) - \int_s^{t+h} Lf(\theta, X(\theta))\,d\theta. \quad (2.9)$$

Replace $a(s,X(s))$ in (2.4) using (2.9). We obtain

$$X(t+h) = x + a(t+h, X(t+h))h - \int_t^{t+h} (s-t)La(s,X(s))\,ds. \quad (2.10)$$

Continuing in this way we find

$$\begin{aligned} X(t+h) &= x + a(t+h,X(t+h))h - La(t+h,X(t+h))\frac{h^2}{2} + \dots \\ &+ (-1)^{m-1}L^{m-1}a(t+h,X(t+h))\frac{h^m}{m!} \\ &+ \int_t^{t+h} (-1)^m \frac{(s-t)^m}{m!} L^m a(s,X(s))\,ds. \end{aligned} \quad (2.11)$$

If we discard in this formula the integral, then we obtain an implicit one-step operation on which we can base a method with mth order of accuracy.

Using a simple trick we can obtain a whole class of implicit methods. We illustrate this trick by deriving a class of implicit methods of second order of accuracy.

We write (2.4) in the following form, introducing a parameter α:

$$X(t+h) = x + \alpha \int_t^{t+h} a(s, X(s))\, ds + (1-\alpha) \int_t^{t+h} a(s, X(s))\, ds. \tag{2.12}$$

Now replace $a(s, X(s))$ in the integral at α by (2.3), and in the integral at $(1-\alpha)$ by (2.9). We obtain

$$\begin{aligned} X(t+h) = x &+ \alpha a(t,x)h + (1-\alpha)a(t+h, X(t+h))h \\ &+ \int_t^{t+h} (t+\alpha h - s) La(s, X(s))\, ds. \end{aligned} \tag{2.13}$$

Writing the integral in (2.13) as the sum of two integrals with coefficients β and $1-\beta$, and using the formula

$$La(s, X(s)) = La(t,x) + O(h)$$

in the first integral and the formula

$$La(s, X(s)) = La(t+h, X(t+h)) + O(h)$$

in the second integral, we obtain

$$\begin{aligned} \overline{X}_{t,x}(t+h) = x + \alpha a(t,x)h + (1-\alpha)a(t+h, \overline{X}_{t,x}(t+h))h + \beta(2\alpha-1)La(t,x)\frac{h^2}{2} \\ + (1-\beta)(2\alpha-1)La(t+h, \overline{X}_{t,x}(t+h))\frac{h^2}{2}. \end{aligned} \tag{2.14}$$

This is an implicit *one-step approximation*, and using it we can construct a two-parameter family of implicit methods of second order of accuracy:

$$\begin{aligned} X_{k+1} = X_k &+ \alpha a_k h + (1-\alpha)a_{k+1}h \\ &+ \beta(2\alpha-1)(La)_k\frac{h^2}{2} + (1-\beta)(2\alpha-1)(La)_{k+1}\frac{h^2}{2}. \end{aligned} \tag{2.15}$$

In (2.15) the functions with index k are computed at (t_k, X_k), while those with index $k+1$ are computed at (t_{k+1}, X_{k+1}).

2.2. Expansion of the solution of a system of stochastic differential equations (Wagner–Platen expansion). Let $X_{t,x}(s) = X(s)$ be the solution of the system (1.1), and let $f(t,x)$ be a sufficiently smooth (scalar or vector) function. By Itô's formula we have for $t_0 \le t \le \theta \le t_0 + T$:

$$f(\theta, X(\theta)) = f(t,x) + \sum_{r=1}^{q} \int_t^{\theta} \Lambda_r f(\theta_1, X(\theta_1))\, dw_r(\theta_1) + \int_t^{\theta} Lf(\theta_1, X(\theta_1))\, d\theta_1, \tag{2.16}$$

where the operators Λ_r, $r = 1, \dots, q$, and L are given by:

$$\Lambda_r = \left(\sigma_r, \frac{\partial}{\partial x}\right),$$

$$L = \frac{\partial}{\partial t} + \left(a, \frac{\partial}{\partial x}\right) + \frac{1}{2}\sum_{r=1}^{q}\sum_{i=1}^{n}\sum_{j=1}^{n}\sigma_r^i\sigma_r^j\frac{\partial^2}{\partial x^i \partial x^j}.$$

Formula (2.16) is the analog of formula (2.3).

Apply (2.16) to the functions $\Lambda_r f$ and Lf, and subsequently insert the expressions obtained for $\Lambda_r f(\theta, X(\theta))$ and $Lf(\theta, X(\theta))$ into (2.16). We find

$$\begin{aligned} f(s, X(s)) = f &+ \sum_{r=1}^{q}\Lambda_r f\int_t^s dw_r(\theta) + Lf\int_t^s d\theta \\ &+ \sum_{r=1}^{q}\int_t^s\left(\sum_{s=1}^{q}\int_t^\theta \Lambda_s\Lambda_r f(\theta_1, X(\theta_1))\,dw_s(\theta_1)\right)dw_r(\theta) \\ &+ \sum_{r=1}^{q}\int_t^s\left(\int_t^\theta L\Lambda_r f(\theta_1, X(\theta_1))\,d\theta_1\right)dw_r(\theta) \\ &+ \sum_{r=1}^{q}\int_t^s\left(\int_t^\theta \Lambda_r L f(\theta_1, X(\theta_1))\,dw_r(\theta_1)\right)d\theta \\ &+ \int_t^s\left(\int_t^\theta L^2 f(\theta_1, X(\theta_1))\,dw\theta_1\right)d\theta, \end{aligned} \tag{2.17}$$

where, e.g., $\Lambda_r f$ is computed at (t, x).

Continuing this way we obtain an expansion for $f(t+h, X(t+h))$. As proved in the previous Subsection, in the deterministic situation this expansion is the Taylor expansion in powers of h with remainder of integral type. In the stochastic situation the role of powers is played by random variables of the form (they are independent of $\mathcal{F}_t$)

$$I_{i_1,\dots,i_j}(h) = \int_t^{t+h} dw_{i_j}(\theta)\int_t^\theta dw_{i_{j-1}}(\theta_1)\int_t^{\theta_1}\cdots\int_t^{\theta_{j-2}} dw_{i_1}(\theta_{j-1}), \tag{2.18}$$

where $i_1, \dots, i_j$ take values in the set $\{0, 1, \dots, q\}$, and where $dw_0(\theta_r)$ is understood to mean $d\theta_r$.

It is obvious that $\mathbf{E}I_{i_1,\dots,i_j} = 0$ if at least one $i_k \neq 0$, $k = 1, \dots, j$, while $\mathbf{E}I_{i_1,\dots,i_j} = O(h^j)$ if all $i_k = 0$, $k = 1, \dots, j$. We evaluate $\mathbf{E}(I_{i_1,\dots,i_j})^2$.

LEMMA 2.1. *We have*

$$\left(\mathbf{E}(I_{i_1,\dots,i_j})^2\right)^{1/2} = O\left(h^{\sum_{k=1}^{j}(2-\bar{\imath}_k)/2}\right), \tag{2.19}$$

where

$$\bar{\imath}_k = \begin{cases} 0, & i_k = 0, \\ 1, & i_k \neq 0. \end{cases}$$

In other words, when computing the *order of smallness of the integral* (2.18) we should be guided by the following rule: $d\theta$ contributes one to the order of smallness, and $dw_r(\theta)$, $r = 1, \ldots, q$, contributes one half.

PROOF. Suppose $i_j \neq 0$. Then (putting $t = 0$ in (2.18))

$$\mathbf{E}(I_{i_1,\ldots,i_j})^2 = \int_0^h \mathbf{E}(I_{i_1,\ldots,i_{j-1}}(\theta))^2 \, d\theta. \tag{2.20}$$

If $i_j = 0$, i.e. $dw_{i_j}(\theta) = d\theta$, then

$$\mathbf{E}(I_{i_1,\ldots,i_j})^2 = \mathbf{E}\left(\int_0^h I_{i_1,\ldots,i_{j-1}}(\theta)\, d\theta\right)^2 \leq h \int_0^h \mathbf{E}(I_{i_1,\ldots,i_{j-1}}(\theta))^2 \, d\theta. \tag{2.21}$$

Let $p(i_1, \ldots, i_j)$ denote the order of smallness of $\mathbf{E}(I_{i_1,\ldots,i_{j-1}})^2$. Then formulas (2.20) and (2.21) give the recurrence relation

$$p(i_1, \ldots, i_j) = p(i_1, \ldots, i_{j-1}) + (2 - \bar{\imath}_j),$$

which proves (2.19). □

To clarify the general rule for establishing expansions of the form (2.17) using the integrals (2.18) we give the following formula, which can be obtained by a series of immediate substitutions (we take s equal to $t + h$):

$$\begin{aligned} f(t+h, X(t+h)) = f &+ \sum_{r=1}^q \Lambda_r f \int_t^{t+h} dw_r(\theta) + Lf \int_t^{t+h} d\theta \\ &+ \sum_{r=1}^q \sum_{i=1}^q \Lambda_i \Lambda_r f \int_t^{t+h} dw_r(\theta) \int_t^{\theta} dw_i(\theta_1) \\ &+ \sum_{r=1}^q \sum_{i=1}^q \sum_{s=1}^q \Lambda_s \Lambda_i \Lambda_r f \int_t^{t+h} dw_r(\theta) \int_t^{\theta} dw_i(\theta_1) \int_t^{\theta_1} dw_s(\theta_2) \\ &+ \sum_{r=1}^q \Lambda_r L f \int_t^{t+h} d\theta \int_t^{\theta} dw_r(\theta_1) + \sum_{r=1}^q L\Lambda_r f \int_t^{t+h} dw_r(\theta) \int_t^{\theta} d\theta_1 \\ &+ L^2 f \int_t^{t+h} d\theta \int_t^{\theta} d\theta_1 + \rho, \end{aligned} \tag{2.22}$$

where

$$\rho = \sum_{r=1}^{q}\sum_{i=1}^{q}\sum_{s=1}^{q}\sum_{j=1}^{q}\int_t^{t+h}\left(\int_t^{\theta}\left(\int_t^{\theta_1}\left(\int_t^{\theta_2}\Lambda_j\Lambda_s\Lambda_i\Lambda_r f(\theta_3, X(\theta_3))\times\right.\right.\right.$$

$$\left.\left.\left.\times\, dw_j(\theta_3)\right) dw_s(\theta_2)\right) dw_i(\theta_1)\right) dw_r(\theta)$$

$$+\sum_{r=1}^{q}\sum_{i=1}^{q}\int_t^{t+h}\left(\int_t^{\theta}\left(\int_t^{\theta_1} L\Lambda_i\Lambda_r f(\theta_2, X(\theta_2))\, d\theta_2\right) dw_i(\theta_1)\right) dw_r(\theta)$$

$$+\sum_{r=1}^{q}\sum_{i=1}^{q}\int_t^{t+h}\left(\int_t^{\theta}\left(\int_t^{\theta_1} \Lambda_i L\Lambda_r f(\theta_2, X(\theta_2))\, dw_i(\theta_2)\right) d\theta_1\right) dw_r(\theta)$$

$$+\sum_{r=1}^{q}\sum_{i=1}^{q}\int_t^{t+h}\left(\int_t^{\theta}\left(\int_t^{\theta_1} \Lambda_i\Lambda_r L f(\theta_2, X(\theta_2))\, dw_i(\theta_2)\right) dw_r(\theta_1)\right) d\theta$$

$$+\sum_{r=1}^{q}\sum_{i=1}^{q}\sum_{s=1}^{q}\int_t^{t+h}\left(\int_t^{\theta}\left(\int_t^{\theta_1}\left(\int_t^{\theta_2} L\Lambda_s\Lambda_i\Lambda_r f(\theta_3, X(\theta_3))\, d\theta_3\right)\right.\right.$$

$$\left.\left.\times\, dw_s(\theta_2)\right) dw_i(\theta_1)\right) dw_r(\theta)$$

$$+\sum_{r=1}^{q}\int_t^{t+h}\left(\int_t^{\theta}\left(\int_t^{\theta_1} L^2\Lambda_r f(\theta_2, X(\theta_2))\, d\theta_2\right) d\theta_1\right) dw_r(\theta)$$

$$+\sum_{r=1}^{q}\int_t^{t+h}\left(\int_t^{\theta}\left(\int_t^{\theta_1} L\Lambda_r L f(\theta_2, X(\theta_2))\, d\theta_2\right) dw_r(\theta_1)\right) d\theta$$

$$+\sum_{r=1}^{q}\int_t^{t+h}\left(\int_t^{\theta}\left(\int_t^{\theta_1} \Lambda_r L^2 f(\theta_2, X(\theta_2))\, dw_r(\theta_2)\right) d\theta_1\right) d\theta$$

$$+\int_t^{t+h}\left(\int_t^{\theta}\left(\int_t^{\theta_1} L^3 f(\theta_2, X(\theta_2))\, d\theta_2\right) d\theta_1\right) d\theta. \tag{2.23}$$

The righthand side of (2.22) consists of a term of zero order of smallness (f), of terms of order of smallness 1/2 (they make up the sum of all possible integrals of the form (2.18) of order 1/2 with corresponding coefficients; each of these terms is $\Lambda_r f \int_t^{t+h} dw_r(\theta)$), of terms of order of smallness 1 (making up the sum of all possible integrals of the form (2.18) of order 1 with corresponding coefficients; here the terms are of two kinds: $Lf \int_t^{t+h} d\theta$ and $\Lambda_s\Lambda_r f \int_t^{t+h} dw_r(\theta) \int_t^{\theta} dw_s(\theta_1)$), of all possible terms of order of smallness 3/2, of one term of order of smallness 2 ($L^2 f \int_t^{t+h} d\theta \int_t^{\theta} d\theta_1$), and of the *remainder* ρ. It is easy to see that the coefficient at the integral$I_{i_1,\dots,i_j}$ (whose order of smallness is $\sum_{k=1}^{j}(2-\bar{\imath}_k)/2$ by Lemma 2.1) is equal to $\Lambda_{i_j}\dots\Lambda_{i_1}$, where Λ_0

is taken to mean L. So, we can obtain an expansion containing all terms up to and including some half-integral order of smallness, as well as an expansion up to and including some integral order of smallness. In (2.22) (discarding ρ) all terms of order of smallness $3/2$ have been included, as well as one term of order of smallness two. This term is characteristic in that the integral in it does not involve Wiener processes, and so its mathematical expectation is not equal to zero (of course, if $L^2 f \neq 0$). The term $L^2 f \int_t^{t+h} d\theta \int_t^\theta d\theta_1$ has been included in the main part of (2.22) for convenience of reference; the true reason for its inclusion will be revealed later.

LEMMA 2.2. *Suppose*

$$\left|\Lambda_{i_j} \dots \Lambda_{i_1} f(t,x)\right| \leq K\left(1+|x|^2\right)^{1/2}. \tag{2.24}$$

Then the quantity

$$I_{i_1,\dots,i_j}(f,h) = \int_t^{t+h} dw_{i_j}(\theta) \int_t^\theta dw_{i_{j-1}}(\theta_1) \int_t^{\theta_1} \dots$$
$$\dots \int_t^{\theta_{j-2}} \Lambda_{i_j} \dots \Lambda_{i_1} f(\theta_{j-1}, X(\theta_{j-1}))\, dw_{i_1}(\theta_{j-1}) \tag{2.25}$$

satisfies the inequality

$$\mathbf{E}\left|I_{i_1,\dots,i_j}(f,h)\right|^2 \leq K\left(1+\mathbf{E}|X(t)|^2\right) h^{\sum_{k=1}^j (2-\bar{i}_k)}, \tag{2.26}$$

i.e., in particular, its order of smallness is the same as that of $I_{i_1,\dots,i_j}(h)$. Furthermore, if at least one index i_k, $k=1,\dots,q$, is not equal to zero, then

$$\mathbf{E} I_{i_1,\dots,i_j}(f,h) = 0, \qquad \sum_{k=1}^j i_k^2 \neq 0. \tag{2.27}$$

The proof of this Lemma does, in essence, not differ from that of Lemma 2.1. We only have to estimate $\mathbf{E}\left|\Lambda_{i_j} \dots \Lambda_{i_1} f(\theta_{j-1}, X(\theta_{j-1}))\right|^2$ after the last step. Using (2.24) we obtain

$$\mathbf{E}\left|\Lambda_{i_j} \dots \Lambda_{i_1} f(\theta_{j-1}, X(\theta_{j-1}))\right|^2 \leq K\left(1+\mathbf{E}|X(\theta_{j-1})|^2\right),$$

whence follows (2.26) in view of the inequality $\mathbf{E}|X(\theta_{j-1})|^2 \leq K\mathbf{E}|X(t)|^2$ for $t < \theta_{j-1}$.

This Lemma implies that each term in the remainder ρ has order of smallness at most two. Moreover, the mathematical expectation of all terms of orders of smallness 2 and $5/2$ from ρ vanishes by (2.27). So, $|E\rho| = O(h^3)$. Of course, this is true if all integrands in ρ satisfy, e.g., (2.24).

We now take in (2.22) and (2.23) x instead of $f(t,x)$. Note that in this case $\Lambda_r f = \sigma_r$, $Lf = a$. Therefore

$$\begin{aligned} X_{t,x}(t+h) = x &+ \sum_{r=1}^{q} \sigma_r \int_t^{t+h} dw_r(\theta) + ah + \sum_{r=1}^{q}\sum_{i=1}^{q} \Lambda_i \sigma_r \int_t^{t+h} (w_i(\theta) - w_i(t))\, dw_r(\theta) \\ &+ \sum_{r=1}^{q} L\sigma_r \int_t^{t+h} (\theta - t) dw_r(\theta) + \sum_{r=1}^{q} \Lambda_r a \int_t^{t+h} (w_r(\theta) - w_r(t))\, d\theta \\ &+ \sum_{r=1}^{q}\sum_{i=1}^{q}\sum_{s=1}^{q} \Lambda_s \Lambda_i \sigma_r \int_t^{t+h} \left(\int_t^{\theta} (w_s(\theta_1) - w_s(t))\, dw_i(\theta_1) \right) dw_r(\theta) \\ &+ La\frac{h^2}{2} + \rho. \end{aligned} \tag{2.28}$$

In this formula all coefficients σ_r, a, $\Lambda_i\sigma_r$, $L\sigma_r$, $\Lambda_r a$, $\Lambda_s\Lambda_i\sigma_r$, and La are computed at the point (t, x), while the remainder ρ is equal to

$$\rho = \sum_{r=1}^{q}\sum_{i=1}^{q}\sum_{s=1}^{q}\sum_{j=1}^{q}\int_{t}^{t+h}\left(\int_{t}^{\theta}\left(\int_{t}^{\theta_1}\left(\int_{t}^{\theta_2}\Lambda_j\Lambda_s\Lambda_i\sigma_r(\theta_3, X(\theta_3))\times\right.\right.\right.$$
$$\left.\left.\left.\times\, dw_j(\theta_3)\right) dw_s(\theta_2)\right) dw_i(\theta_1)\right) dw_r(\theta)$$
$$+\sum_{r=1}^{q}\sum_{i=1}^{q}\int_{t}^{t+h}\left(\int_{t}^{\theta}\left(\int_{t}^{\theta_1} L\Lambda_i\sigma_r(\theta_2, X(\theta_2))\, d\theta_2\right) dw_i(\theta_1)\right) dw_r(\theta)$$
$$+\sum_{r=1}^{q}\sum_{i=1}^{q}\int_{t}^{t+h}\left(\int_{t}^{\theta}\left(\int_{t}^{\theta_1} \Lambda_i L\sigma_r(\theta_2, X(\theta_2))\, dw_i(\theta_2)\right) d\theta_1\right) dw_r(\theta)$$
$$+\sum_{r=1}^{q}\sum_{i=1}^{q}\int_{t}^{t+h}\left(\int_{t}^{\theta}\left(\int_{t}^{\theta_1} \Lambda_i\Lambda_r a(\theta_2, X(\theta_2))\, dw_i(\theta_2)\right) dw_r(\theta_1)\right) d\theta$$
$$+\sum_{r=1}^{q}\sum_{i=1}^{q}\sum_{s=1}^{q}\int_{t}^{t+h}\left(\int_{t}^{\theta}\left(\int_{t}^{\theta_1}\left(\int_{t}^{\theta_2} L\Lambda_s\Lambda_i\sigma_r(\theta_3, X(\theta_3))\, d\theta_3\right)\right.\right.$$
$$\left.\left.\times\, dw_s(\theta_2)\right) dw_i(\theta_1)\right) dw_r(\theta)$$
$$+\sum_{r=1}^{q}\int_{t}^{t+h}\left(\int_{t}^{\theta}\left(\int_{t}^{\theta_1} L^2\sigma_r(\theta_2, X(\theta_2))\, d\theta_2\right) d\theta_1\right) dw_r(\theta)$$
$$+\sum_{r=1}^{q}\int_{t}^{t+h}\left(\int_{t}^{\theta}\left(\int_{t}^{\theta_1} L\Lambda_r a(\theta_2, X(\theta_2))\, d\theta_2\right) dw_r(\theta_1)\right) d\theta$$
$$+\sum_{r=1}^{q}\int_{t}^{t+h}\left(\int_{t}^{\theta}\left(\int_{t}^{\theta_1} \Lambda_r La(\theta_2, X(\theta_2))\, dw_r(\theta_2)\right) d\theta_1\right) d\theta$$
$$+\int_{t}^{t+h}\left(\int_{t}^{\theta}\left(\int_{t}^{\theta_1} L^2 a(\theta_2, X(\theta_2))\, d\theta_2\right) d\theta_1\right) d\theta. \tag{2.29}$$

In relation with the formulas (2.28), (2.29) we consider the following *one-step ap-*

proximations:

$$\overline{X}^{(1)}_{t,x}(t+h) = x + \sum_{r=1}^{q} \sigma_r (w_r(t+h) - w_r(t)), \tag{2.30}$$

$$\overline{X}^{(2)}_{t,x}(t+h) = \overline{X}^{(1)}_{t,x}(t+h) + ah, \tag{2.31}$$

$$\overline{X}^{(3)}_{t,x}(t+h) = \overline{X}^{(2)}_{t,x}(t+h) + \sum_{r=1}^{q}\sum_{i=1}^{q} \Lambda_i \sigma_r \int_t^{t+h} (w_i(\theta) - w_i(t))\, dw_r(\theta), \tag{2.32}$$

$$\begin{aligned} \overline{X}^{(4)}_{t,x}(t+h) = \overline{X}^{(3)}_{t,x}(t+h) &+ \sum_{r=1}^{q} L\sigma_r \int_t^{t+h} (\theta - t)\, dw_r(\theta) \\ &+ \sum_{r=1}^{q} \Lambda_r a \int_t^{t+h} (w_r(\theta) - w_r(t))\, d\theta \\ &+ \sum_{r=1}^{q}\sum_{i=1}^{q}\sum_{s=1}^{q} \Lambda_s \Lambda_i \sigma_r \int_t^{t+h} \left(\int_t^{\theta} (w_s(\theta_1) - w_s(t))\, dw_i(\theta_1) \right) dw_r(\theta), \end{aligned} \tag{2.33}$$

$$\overline{X}^{(5)}_{t,x}(t+h) = \overline{X}^{(4)}_{t,x}(t+h) + La\frac{h^2}{2}. \tag{2.34}$$

To each of these approximations we associate the error $X_{t,x}(t+h) - \overline{X}^{(i)}_{t,x}(t+h) = \rho^{(i)}$, $i = 1, \dots, 5$. Using Lemma 2.1 and Lemma 2.2 it can be readily shown that (under the condition (2.24) on the respective functions)

$$\left|\mathbf{E}\rho^{(1)}\right| = O(h), \qquad \left|\mathbf{E}\rho^{(1)}\right|^2 = O(h^2),$$

i.e. $p_1 = 1$, $p_2 = 1$. Therefore, as p_2 such that the conditions of Theorem 1.1 hold (more precisely, the condition $p_1 \geq p_2 + 1/2$) we can take $p_2 = 1/2$. On the basis of Theorem 1.1 convergence is not guaranteed and, as already noted earlier, the method (2.30) clearly does not converge.

For $\rho^{(2)}$ we have the following identities:

$$\left|\mathbf{E}\rho^{(2)}\right| = O(h^2), \qquad \mathbf{E}\left|\rho^{(2)}\right|^2 = O(h^2),$$

i.e. $p_1 = 2$, $p_2 = 1$. Therefore, the second method (Euler's method) has order of convergence equal to $1/2$.

For $\rho^{(3)}$ we have:

$$\left|\mathbf{E}\rho^{(3)}\right| = O(h^2), \qquad \mathbf{E}\left|\rho^{(3)}\right|^2 = O(h^3),$$

i.e. $p_1 = 2$, $p_2 = 3/2$. Therefore, the third method has order of convergence equal to 1.

For $\rho^{(4)}$ we have:

$$\left|\mathbf{E}\rho^{(4)}\right| = O(h^2), \qquad \mathbf{E}\left|\rho^{(4)}\right|^2 = O(h^4),$$

i.e. $p_1 = 2$, $p_2 = 2$. But to satisfy the conditions of Theorem 1.1 we have to put $p_2 = 3/2$. As a result we see that the fourth method also has order of convergence equal to 1.

Finally, for $\rho^{(5)}$ we have:

$$\left|\mathbf{E}\rho^{(5)}\right| = O(h^3), \qquad \mathbf{E}\left|\rho^{(5)}\right|^2 = O(h^4),$$

i.e. $p_1 = 3$, $p_2 = 2$. Therefore, the fifth method has order of convergence equal to $3/2$.

These examples readily give support to and enable the formulation of a basic result. Indeed, suppose that an expansion of the type under consideration includes all terms of order m. Then the remainder ρ includes terms of half-integral order $m + 1/2$ and of integral order $m + 1$. Since the mathematical expectation of any term of half-integral order vanishes, we have $|\mathbf{E}\rho| = O(h^{m+1})$, i.e. $p_1 = m + 1$. At the same time $\mathbf{E}\rho^2 = O(h^{2m+1})$, i.e. $p_2 = m + 1/2$. By Theorem 1.1 the order of accuracy of such a method is m. On the other hand, if the expansion includes only all terms of half-integral order $m+1/2$, then among the terms of order $m+1$ in the remainder there is, in general, one term having nonzero mathematical expectation; to be precise, such is $\int_t^{t+h} d\theta \int_t^{\theta} d\theta_1 \cdots \int_t^{\theta_{m-1}} L^m a(\theta_m, X(\theta_m))\, d\theta_m$. So, $|\mathbf{E}\rho| = O(h^{m+1})$, $|\mathbf{E}\rho^2|^{1/2} = O(h^{m+1})$. Hence Theorem 1.1 can be applied with only $p_2 = m + 1/2$. The order of accuracy of such a method is again m. Thus, if we add to all terms of order at most m the terms of order $m + 1/2$, then the order of accuracy of the method does not increase. However, if we add to all terms of order at most $m + 1/2$ only the single term of order $m + 1$ referred to above, then the order of accuracy of the method increases by $1/2$. In fact, the mathematical expectation of all remaining terms of order $m + 1$ is zero, and so $|\mathbf{E}\rho| = O(h^{m+2})$, i.e. $p_1 = m + 2$, $p_2 = m + 1$, and $p = m + 1/2$.

Thus, the following theorem holds.

THEOREM 2.1 (E. PLATEN [36]). *Suppose that $\overline{X}_{t,x}(t + h)$ includes all terms of the form $\Lambda_{i_1} \dots \Lambda_{i_j} f I_{i_1,\dots,i_j}$, where $f \equiv x$, up to order m inclusive. Suppose that all functions $\Lambda_{i_1} \dots \Lambda_{i_j} f(t,x)$, where $f \equiv x$, $\sum_{k=1}^{j}(2 - \bar{\imath}_k)/2 \le m + 1$, satisfy the inequality* (2.24). *Then the mean-square order of accuracy of the method based on this approximation is equal to m.*

Suppose that $\overline{X}_{t,x}(t+h)$ includes all terms of the form $\Lambda_{i_1} \dots \Lambda_{i_j} f I_{i_1,\dots,i_j}$, where $f \equiv x$, up to order $m+1/2$ inclusive, as well as the term $L^m a \int_t^{t+h} d\theta \int_t^{\theta} d\theta_1 \cdots \int_t^{\theta_{m-1}} d\theta_m = L^m a h^{m+1}/(m+1)!$. Suppose that all functions $\Lambda_{i_1} \dots \Lambda_{i_j} f(t,x)$, where $f \equiv x$, $\sum_{k=1}^{j}(2-\bar{\imath}_k)/2 \le m + 2$, satisfy the inequality (2.24). *Then the mean-square order of accuracy of the method based on this approximation is equal to $m + 1/2$.*

REMARK 2.1. Not all expressions of order $5/2$ occur in the remainder ρ (see (2.29)). For example, the expression with integrand $\Lambda_s L \Lambda_i \sigma_r$ does not occur in it. Therefore, the requirement in the theorem that all corresponding functions satisfy the inequality (2.24) can be weakened a bit.

EXAMPLE 2.1. Consider the system of stochastic differential equations

$$dX = A(t)X\, dt + \sum_{r=1}^{q} B_r(t)X\, dw_r(t), \qquad t_0 \le t \le t_0 + T. \tag{2.35}$$

Here, $A(t)$ and $B_r(t)$ are $(n\times n)$-matrices with entries that are smooth on $[t_0, t_0+T]$, and $a(t,x) = A(t)x$, $\sigma_r(t,x) = B_r(t)x$. Therefore, for any construction of a method of any order of accuracy the conditions of Theorem 2.1 hold. We take the method of first order of accuracy given by the approximation (2.32). Since

$$\Lambda_i\sigma_r(t,x) = \left(B_i(t)x, \frac{\partial}{\partial x}\right) B_r(t)x = B_r(t)B_i(t)x,$$

this method has the form

$$X_{k+1} = X_k + \sum_{r=1}^{q} B_r(t_k)X_k\Delta_k w_k(h) + A(t_k)X_k h$$
$$+ \sum_{r=1}^{q}\sum_{i=1}^{q} B_r(t_k)B_i(t_k)X_k \int_{t_k}^{t_k+h} (w_i(\theta) - w_i(t_k))\, dw_r(\theta). \tag{2.36}$$

It is easy to verify the formula

$$\int_{t_k}^{t_k+h} (w_i(\theta) - w_i(t_k))\, dw_r(\theta) = \Delta_k w_i(h)\Delta_k w_r(h) - \int_{t_k}^{t_k+h} (w_r(\theta) - w_r(t_k))\, dw_i(\theta).$$

If all $B_r(t)$ commute, then for $i \neq r$ we have

$$B_rB_iX \int_{t_k}^{t_k+h} (w_i(\theta) - w_i(t_k))\, dw_r(\theta) + B_iB_rX \int_{t_k}^{t_k+h} (w_r(\theta) - w_r(t_k))\, dw_i(\theta)$$
$$= B_iB_rX\Delta_k w_i(h)\Delta_k w_r(h).$$

Formula (2.36) now takes the form

$$X_{k+1} = X_k + \sum_{r=1}^{q} B_r(t_k)X_k\Delta_k w_k(h) + A(t_k)X_k h + \frac{1}{2}\sum_{i=1}^{q} B_i^2(t_k)X_k\left(\Delta_k^2 w_i(h) - h\right)$$
$$+ \sum_{r=2}^{q}\sum_{i=1}^{r-1} B_i(t_k)B_r(t_k)X_k\Delta_k w_i(h)\Delta_k w_r(h). \tag{2.37}$$

Thus, in the *commutative situation* we can construct a method of first order of accuracy by modeling only the increments of the Wiener processes.

We also note that in the nonlinear case (2.32) can be simplified in a similar manner, if

$$\Lambda_i\sigma_r(t,x) = \Lambda_r\sigma_i(t,x).$$

2.3. Construction of implicit methods. Here we consider, next to (2.16), the formula ($t \le \theta_1 \le \theta$)

$$Lf(\theta_1, X(\theta_1)) = Lf(\theta, X(\theta)) - \sum_{r=1}^{q} \int_{\theta_1}^{\theta} \Lambda_r Lf(\theta_2, X(\theta_2))\, dw_r(\theta_2) - \int_{\theta_1}^{\theta} L^2 f(\theta_2, X(\theta_2))\, d\theta_2. \tag{2.38}$$

As in the deterministic case, we substitute (2.38) into (2.16), putting $\theta = t+h$, and obtain

$$f(t+h, X(t+h)) = f(t,x) + \sum_{r=1}^{q} \int_t^{t+h} \Lambda_r f(\theta_1, X(\theta_1))\, dw_r(\theta_1) + Lf(t+h, X(t+h))h - \sum_{r=1}^{q} \int_t^{t+h} \left(\int_{\theta_1}^{t+h} \Lambda_r Lf(\theta_2, X(\theta_2))\, dw_r(\theta_2) \right) d\theta_1 - \int_t^{t+h} \left(\int_{\theta_1}^{t+h} L^2 f(\theta_2, X(\theta_2))\, d\theta_2 \right) d\theta_1. \tag{2.39}$$

To use a formula of the form (2.38) to represent $\Lambda_r f(\theta_1, X(\theta_1))$ with subsequent substitution into (2.39) is rather unwise. In fact, although the function $\Lambda_r f(\theta_1, X(\theta_1))$ does not depend itself on the future, all terms in its representation of the form (2.38) do depend on the future and, e.g., the integral $\int_t^{t+h} \left(\int_{\theta_1}^{t+h} L\Lambda_r f(\theta_2, X(\theta_2))\, d\theta_2 \right) dw_r(\theta_1)$ does not make sense without some additional clarification. This complication can be overcome using the following trick. Consider, e.g., formula (2.22) and represent the coefficient $\Lambda_r f$ in it as follows:

$$\Lambda_r f(t,x) = \Lambda_r f(t+h, X(t+h)) - \sum_{i=1}^{q} \int_t^{t+h} \Lambda_i \Lambda_r f(\theta, X(\theta))\, dw_i(\theta) - \int_t^{t+h} L\Lambda_r f(\theta, X(\theta))\, d\theta. \tag{2.40}$$

After substituting (2.40) into (2.22), the righthand side shows a dependency on $X(t+h)$ as a factor not of h, as in (2.39), but of $\Delta w_r(h)$. The appearance of this implicitness may lead to a method known to be inapplicable. We clarify this by the example of the equation

$$dX = aX\, dt + \sigma X\, dw.$$

We have

$$X_{t,x}(t+h) = x + \sigma x \Delta w(h) + axh + \rho,$$

where $\mathbf{E}\rho = O(h^2)$, $\mathbf{E}\rho^2 = O(h^2)$.

Further, as in (2.40), σx can be written as

$$\sigma x = \sigma X(t+h) - \int\limits_t^{t+h} \sigma X(\theta)\, dw(\theta) - \int\limits_t^{t+h} aX(\theta)\, d\theta.$$

As a result we find

$$X(t+h) = x + \sigma X(t+h)\Delta w(h) + axh + \rho_1. \tag{2.41}$$

Discarding ρ_1 (at this moment we will not be concerned with justifying this) we obtain the method

$$X_{k+1} = X_k + \sigma X_{k+1}\Delta w(h) + aX_k h, \tag{2.42}$$

which, similar to the method (1.52) in Example 1.2, gives X_{k+1} with infinite second moment. Generally, if at the righthand side of an equation containing X_{k+1} we have factors $\Delta w(h)$, i.e. quantities taking arbitrary large values, then the solvability with respect to X_{k+1} of the relations obtained in such situations is questionable. At the same time, if the expressions containing X_{k+1} involve a factor h with positive power, then for sufficiently small h the solvability is guaranteed, given natural assumptions. Precisely therefore a certain caution is necessary when introducing implicitness because of expressions occurring in stochastic integrals. Of course, it can be hoped that by introducing implicitness only in expressions occurring in nonstochastic integrals we will succeed in adding corresponding stability to the methods; just for this reason we construct implicit methods. Again we return to (2.22), and write the coefficient Lf as

$$\begin{aligned} Lf &= Lf(t+h, X(t+h)) - \sum_{r=1}^{q} \int\limits_t^{t+h} \Lambda_r Lf(\theta, X(\theta))\, dw_r(\theta) - \int\limits_t^{t+h} L^2 f(\theta, X(\theta))\, d\theta \\ &= Lf(t+h, X(t+h)) - \sum_{r=1}^{q} \Lambda_r Lf \int\limits_t^{t+h} dw_r(\theta) - L^2 f \int\limits_t^{t+h} d\theta + \rho_1, \end{aligned} \tag{2.43}$$

where, as can be readily shown,

$$|\mathbf{E}\rho_1 h| \le K\left(1+|x|^2\right)^{1/2} h^3, \qquad \mathbf{E}\rho_1^2 h^2 \le K\left(1+|x|^2\right)^{1/2} h^4.$$

The other coefficients of (2.22) can be treated in the same manner (e.g., the reasoning above applies just as well to the coefficient $L^2 f$). Moreover, similar transformations can be repeatedly performed, and also over individual terms containing, say, Lf, since a formula of the type (2.43) can be written down for any smooth function. As a result we obtain a large amount of distinct representations for $f(t+h, X(t+h))$ using the integrals $I_{i_1,\dots,i_j}$ or using products of such integrals with coefficients depending on the points (t,x) and $(t+h, X(t+h))$. As in the deterministic situation, the amount of such representations can be increased by considering splittings of, e.g., the term $Lf \int_t^{t+h} d\theta$ into the sum of two integrals with coefficients α and $(1-\alpha)$ such that the term $\alpha Lf \int_t^{t+h} d\theta$ remains unchanged while in the term $(1-\alpha)Lf \int_t^{t+h} d\theta$ we replace Lf by (2.43). In §3 we will give and use concrete implicit methods obtained on the basis of such representations. Note that the wish to construct a large amount of methods, both implicit and explicit, is related to the fact that distinct methods have various properties as regards accuracy, stability, time and labour consumation, etc.

3. Explicit and implicit methods of order 3/2 for systems with additive noises

The majority of results obtained in this Section can without difficulty be generalised to the case of general systems and methods of higher order of accuracy. The choice of the form of the systems and methods indicated in the title of this Section is related to the wish to avoid unnecessary complications in the exposition of the subsequent constructions and methods of investigation. Moreover, while systems with additive noises are convenient for demonstrating the construction of various approximate methods, they are also of great independent interest. From the point of view of numerical integration, their distinguishing mark is the absence of random variables of the form I_{i_1,i_2} and I_{i_1,i_2,i_3}, $i_1, i_2, i_3 \neq 0$, in the Taylor-type expansions, and therefore we are able to construct various constructive (with respect to modeling of random variables) methods with order of accuracy reaching 3/2.

3.1. Explicit methods based on Taylor-type expansion. Consider the following *system of stochastic differential equations with additive noises*:

$$dX = a(t, X)\,dt + \sum_{r=1}^{q} \sigma_r(t)\,dw_r(t). \tag{3.1}$$

We use the formulas (2.28) and (2.29). We have

$$\Lambda_i \sigma_r = 0, \qquad L\sigma_r = \frac{d\sigma_r}{dt}(t), \qquad \Lambda_r a = \left(\sigma_r, \frac{\partial}{\partial x}\right) a(t,x), \qquad \Lambda_s \Lambda_i \sigma_r = 0,$$

$$La = \frac{\partial a}{\partial t}(t,x) + \left(a, \frac{\partial}{\partial x}\right) a(t,x) + \frac{1}{2} \sum_{r=1}^{q} \sum_{i=1}^{n} \sum_{j=1}^{n} \sigma_r^i \sigma_r^j \frac{\partial^2 a}{\partial x^i \partial x^j}(t,x).$$

By (2.28), we can write down the following numerical integration formula:

$$X_{k+1} = X_k + \sum_{r=1}^{q} \sigma_{r_k} \Delta_k w_r(h) + a_k h + \sum_{r=1}^{q} (\Lambda_r a)_k \int_{t_k}^{t_{k+1}} (w_r(\theta) - w_r(t_k))\,d\theta$$

$$+ \sum_{r=1}^{q} \sigma'_{r_k} \int_{t_k}^{t_{k+1}} (\theta - t_k)\,dw_r(\theta) + (La)_k \frac{h^2}{2}. \tag{3.2}$$

Suppose the functions $a(t,x)$ and $\sigma_r(t)$ satisfy for $t_0 \leq t \leq t_0 + T$, $x \in \mathbb{R}^n$ the following conditions (to be called *conditions A* in the sequel):

- The function a and all its first- and second-order partial derivatives, as well as the partial derivatives $\partial^3 a / \partial t \partial x^i \partial x^j$, $\partial^3 a / \partial x^i \partial x^j \partial x^k$, and $\partial^4 a / \partial x^i \partial x^j \partial x^k \partial x^l$, are continuous;
- The functions $\sigma_r(t)$ are twice continuously differentiable.

Suppose that the first-order partial derivatives of a with respect to x are uniformly bounded (so that a Lipschitz condition is satisfied), while its remaining partial derivatives listed above, regarded as functions of x, grow at most as a linear function of $|x|$ as $|x| \to \infty$. Then (see Remark 2.1 and (2.29)) Theorem 2.1 holds. Thus, the method (3.2) has order of accuracy equal to 3/2.

In (3.2) the random variables $\Delta_k w_r(h)$,

$$\int_{t_k}^{t_{k+1}} (w_r(\theta) - w_r(t_k))\, d\theta, \qquad \int_{t_k}^{t_{k+1}} (\theta - t_k)\, dw_r(\theta), \qquad r = 1, \dots, q,$$

play a role. Since

$$\int_{t_k}^{t_{k+1}} (\theta - t_k)\, dw_r(\theta) = h\Delta_k w_r(h) - \int_{t_k}^{t_{k+1}} (w_r(\theta) - w_r(t_k))\, d\theta, \tag{3.3}$$

to use (3.2) at the $(k+1)$st step it suffices to model the random variables $\Delta_k w_r(h)$ and $\int_{t_k}^{t_{k+1}} (w_r(\theta) - w_r(t_k))\, d\theta$, $r = 1, \dots, q$. Each of these has a Gaussian distribution. We can immediately compute that

$$\mathbf{E}\left[\Delta_k w_r(h)\left(\int_{t_k}^{t_{k+1}} (w_r(\theta) - w_r(t_k))\, d\theta - \frac{1}{2}h\Delta_k w_r(h)\right)\right] = 0, \tag{3.4}$$

$$\mathbf{E}\,(\Delta_k w_r(h))^2 = h, \tag{3.5}$$

$$\mathbf{E}\left(\int_{t_k}^{t_{k+1}} (w_r(\theta) - w_r(t_k))\, d\theta - \frac{1}{2}h\Delta_k w_r(h)\right)^2 = \frac{1}{12}h^3. \tag{3.6}$$

In fact,

$$\mathbf{E}\left(\Delta_k w_r(h) \int_{t_k}^{t_{k+1}} (w_r(\theta) - w_r(t_k))\, d\theta\right)$$
$$= \int_{t_k}^{t_{k+1}} \mathbf{E}\,(\Delta_k w_r(h)\,(w_r(\theta) - w_r(t_k)))\, d\theta = \int_{t_k}^{t_{k+1}} (\theta - t_k)\, d\theta = \frac{h^2}{2}, \tag{3.7}$$

which implies (3.4). By (3.7), the proof of (3.6) reduces to the computation

$$\mathbf{E}\left(\int_{t_k}^{t_{k+1}} (w_r(\theta) - w_r(t_k))\, d\theta\right)^2 = 2\mathbf{E}\int_{t_k}^{t_{k+1}}\left(\int_{t_k}^{\theta}(w_r(s) - w_r(t_k))\, ds\, (w_r(\theta) - w_r(t_k))\right) d\theta$$

$$= 2\int_{t_k}^{t_{k+1}}\int_{t_k}^{\theta}\mathbf{E}\left[(w_r(s) - w_r(t_k))\,(w_r(\theta) - w_r(t_k))\right] ds\, d\theta$$

$$= 2\int_{t_k}^{t_{k+1}}\int_{t_k}^{\theta}(s - t_k)\, ds\, d\theta = \frac{h^3}{3}. \tag{3.8}$$

Introduce the following independent $N(0,1)$-distributed random variables ξ_{rk} and η_{rk}:

$$\xi_{rk} = h^{-1/2}\Delta_k w_r(h),$$
$$\eta_{rk} = \sqrt{12}h^{-3/2}\left(\int_{t_k}^{t_{k+1}} (w_r(\theta) - w_r(t_k))\, d\theta - \frac{1}{2}h\Delta_k w_r(h)\right). \tag{3.9}$$

Using these random variables we obtain

$$\Delta_k w_r(h) = h^{1/2}\xi_{rk},$$
$$\int_{t_k}^{t_{k+1}} (w_r(\theta) - w_r(t_k))\, d\theta = h^{3/2}\left(\frac{1}{2}\xi_{rk} + \frac{1}{\sqrt{12}}\eta_{rk}\right). \tag{3.10}$$

As a result, formula (3.2) takes the following concrete form:

$$X_{k+1} = X_k + \sum_{r=1}^{q}\sigma_r(t_k)\xi_{rk}h^{1/2} + a(t_k, X_k)h$$
$$+ \sum_{r=1}^{q}\Lambda_r a(t_k, X_k)\left(\frac{1}{2}\xi_{rk} + \frac{1}{\sqrt{12}}\eta_{rk}\right)h^{3/2}$$
$$+ \sum_{r=1}^{q}\frac{d\sigma_r}{dt}(t_k)\left(\frac{1}{2}\xi_{rk} - \frac{1}{\sqrt{12}}\eta_{rk}\right)h^{3/2} + La(t_k, X_k)\frac{h^2}{2}. \tag{3.11}$$

The terms in (3.2) and (3.11) containing $(d\sigma_r/dt)(t_k)$ appear because of the approximate representation of the integral $\int_{t_k}^{t_{k+1}}\sigma_r(\theta)\, dw_r(\theta)$ in the form

$$\int_{t_k}^{t_{k+1}}\sigma_r(\theta)\, dw_r(\theta) = \sigma_r(t_k)\Delta_k w_r(h) + \frac{d\sigma_r}{dt}(t_k)\int_{t_k}^{t_{k+1}}(\theta - t_k)\, dw_r(\theta) + \rho, \tag{3.12}$$

where ρ satisfies the relations $\mathbf{E}\rho = 0$, $\mathbf{E}\rho^2 = O(h^5)$.

Moreover, the random variables $\int_{t_k}^{t_{k+1}} \sigma_r(\theta)\, dw_r(\theta)$, $r = 1, \dots, q$, have a Gaussian distribution. If we model the integrals $\int_{t_k}^{t_{k+1}} \sigma_r(\theta)\, dw_r(\theta)$ exactly, then we can thus avoid the computation of $(d\sigma_r/dt)(t_k)$ and drop the requirement that $\sigma_r(t)$ be smooth.

We give the required computations. First, for the system (3.1), instead of (2.28)–(2.29) we write down

$$\begin{aligned} X(t+h) = x + \sum_{r=1}^{q} \int_t^{t+h} \sigma_r(\theta)\, dw_r(\theta) + ah \\ + \sum_{r=1}^{q} \Lambda_r a \int_t^{t+h} (w_r(\theta) - w_r(t))\, d\theta + La\frac{h^2}{2} + \rho, \end{aligned} \tag{3.13}$$

$$\begin{aligned} \rho = \sum_{r=1}^{q}\sum_{i=1}^{q} \int_t^{t+h} \left(\int_t^{\theta} \left(\int_t^{\theta_1} \Lambda_i \Lambda_r a(\theta_2, X(\theta_2))\, dw_i(\theta_2) \right) dw_r(\theta_1) \right) d\theta \\ + \sum_{r=1}^{q} \int_t^{t+h} \left(\int_t^{\theta} \left(\int_t^{\theta_1} L\Lambda_r a(\theta_2, X(\theta_2))\, d\theta_2 \right) dw_r(\theta_1) \right) d\theta \\ + \sum_{r=1}^{q} \int_t^{t+h} \left(\int_t^{\theta} \left(\int_t^{\theta_1} \Lambda_r La(\theta_2, X)(\theta_2))\, dw_r(\theta_2) \right) d\theta_1 \right) d\theta \\ + \int_t^{t+h} \left(\int_t^{\theta} \left(\int_t^{\theta_1} L^2 a(\theta_2, X(\theta_2))\, d\theta_2 \right) d\theta_1 \right) d\theta. \end{aligned} \tag{3.14}$$

Instead of (3.2) we consider the method

$$\begin{aligned} X_{k+1} = X_k + \sum_{r=1}^{q} \int_{t_k}^{t_{k+1}} \sigma_r(\theta)\, dw_r(\theta) + a_k h \\ + \sum_{r=1}^{q} (\Lambda_r a)_k \int_{t_k}^{t_{k+1}} (w_r(\theta) - w_r(t_k))\, d\theta + (La)_k \frac{h^2}{2}. \end{aligned} \tag{3.15}$$

Here we impose on a the same conditions as in the method (3.11), and we impose on σ_r merely the continuity condition. Theorem 2.1 then gives that the order of accuracy of the method (3.15) is 3/2. We will say something on modeling the random variables occurring in the method (3.15). For each (r, k) the collections of random variables

$$\alpha_{rk}^i = \int_{t_k}^{t_{k+1}} \sigma_r^i(\theta)\, dw_r(\theta), \qquad i = 1, \dots, n, \qquad \beta_{rk} = \int_{t_k}^{t_{k+1}} (w_r(\theta) - w_r(t_k))\, d\theta$$

have a Gaussian distribution, while for $(r_1, k_1) \neq (r_2, k_2)$ these collections are independent. If in the method (3.11) it is necessary to model q pairs of normally distributed random variables at each step, then at each step in (3.15) we have to model q normally distributed collections involving $n + 1$ variables in each collection. In this respect the method (3.15) loses substantially from the method (3.11). We give some characteristics of the normally distributed vector with components $\alpha^1_{rk}, \dots, \alpha^n_{rk}, \beta_{rk}$. We have (see (3.3) and (3.8)):

$$\mathbf{E}\alpha^i_{rk} = \mathbf{E}\beta_{rk} = 0, \qquad \mathbf{E}\alpha^i_{rk}\alpha^j_{rk} = \int\limits_{t_k}^{t_{k+1}} \sigma^i_r(\theta)\sigma^j_r(\theta)\,d\theta, \qquad \mathbf{E}\beta^2_{rk} = \frac{h^3}{3},$$

$$\mathbf{E}\alpha^i_{rk}\beta_{rk} = \mathbf{E}\left[\int\limits_{t_k}^{t_{k+1}} \sigma^i_r(\theta)\,dw_r(\theta)\cdot\left(h\int\limits_{t_k}^{t_{k+1}} dw_r(\theta) - \int\limits_{t_k}^{t_{k+1}} (\theta - t_k)\,dw_r(\theta)\right)\right]$$
$$= \int\limits_{t_k}^{t_{k+1}} (t_{k+1} - \theta)\sigma^i_r(\theta)\,d\theta.$$

We gather the results in this Subsection in a Theorem. Moreover, here and below, as a rule we will not precisely indicate the conditions on the coefficients a and σ_r, and will be satisfied by saying that these coefficients satisfy appropriate smoothness and boundedness conditions (conditions of type A). In each instance the conditions can be exactly established without difficulty, basing oneself on Theorem 3.1 and its proof as done earlier with the remainders (2.29) and (3.14).

THEOREM 3.1. *Suppose the coefficients $a(t,x)$ and $\sigma_r(t)$ of* (3.1) *satisfy conditions of type A. Then both the method* (3.11), *in which the independent $N(0,1)$-distributed random variables ξ_{rk} and η_{rk} are defined using the Wiener processes of the system* (3.1) *by* (3.9), *and the method* (3.15) *have mean-square order of accuracy equal to* 3/2.

3.2. Implicit methods based on Taylor-type expansion. Following the advice of Subsection 2.3, we write the term a in (3.13) as a sum $\alpha a + (1-\alpha)a$. In the second term of this sum we replace a by

$$a(t,x) = a(t+h, X(t+h)) - \sum_{r=1}^{q}\int\limits_{t}^{t+h} \Lambda_r a(\theta, X(\theta))\,dw_r(\theta) - \int\limits_{t}^{t+h} La(\theta, X(\theta))\,d\theta$$
$$= a(t+h, X(t+h)) - \sum_{r=1}^{q}\Lambda_r a\int\limits_{t}^{t+h} dw_r(\theta) - La\cdot h + \rho_1, \tag{3.16}$$

where

$$\rho_1 = -\sum_{r=1}^{q}\sum_{i=1}^{q}\int_t^{t+h}\left(\int_t^{\theta}\Lambda_i\Lambda_r a(\theta_1, X(\theta_1))\,dw_i(\theta_1)\right)dw_r(\theta)$$
$$-\sum_{r=1}^{q}\int_t^{t+h}\left(\int_t^{\theta}L\Lambda_r a(\theta_1, X(\theta_1))\,d\theta_1\right)dw_r(\theta)$$
$$-\sum_{r=1}^{q}\int_t^{t+h}\left(\int_t^{\theta}\Lambda_r La(\theta_1, X(\theta_1))\,dw_r(\theta_1)\right)d\theta$$
$$-\int_t^{t+h}\left(\int_t^{\theta}L^2a(\theta_1, X(\theta_1))\,d\theta_1\right)d\theta. \tag{3.17}$$

Substitute (3.16) into (3.13). The relation obtained will contain a term $(2\alpha-1)La\cdot h^2/2$. Again we write La as a sum $\beta La + (1-\beta)La$, and replace in the second term La by

$$La(t,x) = La(t+h, X(t+h)) + \rho_2, \tag{3.18}$$

where

$$\rho_2 = -\int_t^{t+h}L^2a(\theta, X(\theta))\,d\theta - \sum_{r=1}^{q}\int_t^{t+h}\Lambda_r La(\theta, X(\theta))\,dw_r(\theta). \tag{3.19}$$

Combining all these expressions, we obtain

$$X(t+h) = x + \sum_{r=1}^{q}\int_t^{t+h}\sigma_r(\theta)\,dw_r(\theta)$$
$$+\alpha ah + (1-\alpha)a(t+h, X(t+h))h$$
$$-(1-\alpha)h\sum_{r=1}^{q}\Lambda_r a\int_t^{t+h}dw_r(\theta) + \sum_{r=1}^{q}\Lambda_r a\int_t^{t+h}(w_r(\theta)-w_r(t))\,d\theta$$
$$+\beta(2\alpha-1)La\frac{h^2}{2} + (1-\beta)(2\alpha-1)La(t+h, X(t+h))\frac{h^2}{2}$$
$$+\rho + (1-\alpha)\rho_1 h + (2\alpha-1)(1-\beta)\rho_2\frac{h^2}{2}, \tag{3.20}$$

where ρ is defined by (3.14), ρ_1 by (3.17), and ρ_2 by (3.19).

It can be readily seen that $R = \rho + (1-\alpha)\rho_1 h + (2\alpha-1)(1-\beta)\rho_2 h^2/2$ satisfies the conditions (of course, with appropriate conditions on a; the σ_r are assumed to be continuous):

$$|\mathbf{E}R| \le K\left(1+|x|^2\right)^{1/2}h^3, \qquad \left(\mathbf{E}R^2\right)^{1/2} \le K\left(1+|x|^2\right)^{1/2}h^2.$$

If we discard in (3.20) the term R, then we obtain an implicit *one-step approximation* whose realisation requires the modeling of the integrals $\int_t^{t+h}\sigma_r(\theta)\,dw_r(\theta)$. If $\sigma_r''(t)$ exist and are bounded, we can use the representation (3.12). Substituting this into (3.20) we obtain a new remainder, R_1, which, as can be readily seen, satisfies the same inequalities as R (see above). We write the one-step approximation thus obtained as

$$\begin{aligned}
\overline{X}(t+h) &= x + \sum_{r=1}^{q} \sigma_r(t)\,(w_r(t+h) - w_r(t)) \\
&+ \alpha a(t,x)h + (1-\alpha)a(t+h,\overline{X}(t+h))h \\
&- (1-\alpha)h \sum_{r=1}^{q} \Lambda_r a(t,x)\,(w_r(t+h) - w_r(t)) \\
&+ \sum_{r=1}^{q} \Lambda_r a(t,x) \int_t^{t+h} (w_r(\theta) - w_r(t))\,d\theta \\
&+ \sum_{r=1}^{q} \sigma_r'(t)\,(w_r(t+h) - w_r(t))\,h - \int_t^{t+h} (w_r(\theta) - w_r(t))\,d\theta \\
&+ \beta(2\alpha-1)La(t,x)\frac{h^2}{2} + (1-\beta)(2\alpha-1)La(t+h,\overline{X}(t+h))\frac{h^2}{2}.
\end{aligned} \tag{3.21}$$

If we denote the righthand side of (3.21) by $F(\overline{X}(t+h))$, regarding all other variables as parameters, then next to (3.21), which can be written as

$$\overline{X}(t+h) = F(\overline{X}(t+h)), \tag{3.22}$$

we can write down the equation

$$X(t+h) = F(X(t+h)) + R_1. \tag{3.23}$$

The following two-parameter implicit method corresponds to the approximation (3.21):

$$\begin{aligned}
X_{k+1} &= X_k + \sum_{r=1}^{q} \sigma_r(t_k)\xi_{rk}h^{1/2} \\
&+ \alpha a(t_k,X_k)h + (1-\alpha)a(t_{k+1},X_{k+1})h \\
&+ \sum_{r=1}^{q} \Lambda_r a(t_k,X_k)\left(\frac{2\alpha-1}{2}\xi_{rk} + \frac{1}{\sqrt{12}}\eta_{rk}\right)h^{3/2} \\
&+ \sum_{r=1}^{q} \sigma_r'(t_k)\left(\frac{1}{2}\xi_{rk} - \frac{1}{\sqrt{12}}\eta_{rk}\right)h^{3/2} \\
&+ \beta(2\alpha-1)La(t_k,X_k)\frac{h^2}{2} + \beta(2\alpha-1)La(t_{k+1},X_{k+1})\frac{h^2}{2},
\end{aligned} \tag{3.24}$$

where ξ_{rk} and η_{rk} are the same as in the method (3.11).

THEOREM 3.2. *Suppose that the coefficients $a(t,x)$ and $\sigma_r(t)$ of (3.1) satisfy conditions of type A and, moreover, that $a(t,x)$ has bounded third-order derivative with respect to x (so that La satisfies a uniform Lipschitz condition). Then the implicit one-step approximation (3.21) satisfies the conditions of Theorem 1.1 with $p_1 = 3$,*

$p_2 = 2$, and so the order of accuracy of the method (3.24) is equal to 3/2. The order of accuracy of the method based on one-step approximation according to formula (3.20) by discarding R is also equal to 3/2.

PROOF. Compute the difference $X(t+h) - \overline{X}(t+h)$ using (3.21)–(3.23):

$$X(t+h) - \overline{X}(t+h) = (1-\alpha)\left(a(t+h, X(t+h)) - a(t+h, \overline{X}(t+h))\right) h + (1-\beta)(2\alpha - 1) \times \left(La(t+h, X(t+h)) - La(t+h, \overline{X}(t+h))\right)\frac{h^2}{2} + R_1. \tag{3.25}$$

Since a and La satisfy a Lipschitz condition, we have

$$\left|X(t+h) - \overline{X}(t+h)\right| \le |1-\alpha| \cdot h \cdot K \left|X(t+h) - \overline{X}(t+h)\right| + |1-\beta| \cdot |2\alpha - 1| \cdot K\frac{h^2}{2}\left|X(t+h) - \overline{X}(t+h)\right| + |R_1|.$$

Whence, for sufficiently small h,

$$\left|X(t+h) - \overline{X}(t+h)\right| \le 2\,|R_1|,$$

and so (recall that R_1 satisfies inequalities of the same type as R, since $X(t) = \overline{X}(t) = x$),

$$\mathbf{E}\left|X(t+h) - \overline{X}(t+h)\right|^2 \le K\left(1 + |x|^2\right) h^4. \tag{3.26}$$

Further, (3.25) implies

$$\left|\mathbf{E}\left(X(t+h) - \overline{X}(t+h)\right)\right| \le |1-\alpha| \cdot h \cdot K \cdot \mathbf{E}\left|X(t+h) - \overline{X}(t+h)\right| + |1-\beta| \cdot |2\alpha - 1| \cdot K\frac{h^2}{2}\mathbf{E}\left|X(t+h) - \overline{X}(t+h)\right| + |\mathbf{E}R_1|.$$

Whence, since by (3.26)

$$\mathbf{E}\left|X(t+h) - \overline{X}(t+h)\right| \le K\left(1 + |x|^2\right)^{1/2} h^2,$$

we have

$$\left|\mathbf{E}\left(X(t+h) - \overline{X}(t+h)\right)\right| \le K\left(1 + |x|^2\right)^{1/2} h^3. \tag{3.27}$$

The inequalities (3.26), (3.27) and Theorem 1.1 show that the method (3.24) has order of accuracy equal to 3/2. The second part of the Theorem can be proved in a similar manner. □

EXAMPLE 3.1. For $\alpha = 1/2$ the method (3.24) becomes

$$X_{k+1} = X_k + \sum_{r=1}^{q} \sigma_r(t_k)\xi_{rk}h^{1/2} + (a(t_k, X_k) + a(t_{k+1}, X_{k+1}))\frac{h^2}{2}$$
$$+ \sum_{r=1}^{q} \Lambda_r a(t_k, X_k)\frac{1}{\sqrt{12}}\eta_{rk}h^{3/2} + \sum_{r=1}^{q} \sigma_r'(t_k)\left(\frac{1}{2}\xi_{rk} - \frac{1}{\sqrt{12}}\eta_{rk}\right)h^{3/2}. \tag{3.28}$$

In certain respects this method is even simpler than the method (3.11) (it does not contain La). Moreover, as can be seen from the proof of the Theorem, in this case it is not necessary to require La to satisfy a Lipschitz condition.

3.3. Stiff systems of stochastic differential equations with additive noises. A-stability. As is well known one often meets deterministic systems for which the application of explicit methods (e.g., Runge–Kutta methods of various orders of accuracy, Adams methods) requires the use of a very small step h on the whole interval of integration. Here, while on a relatively small interval of rapid change of the solution the choice of h is dictated by interpolation conditions, on a substantially larger part of the interval there is no objective necessity for choosing a very small step, and this small step arises only as a consequence of certain instability properties of the method itself. We clarify this by an example. Consider the two-dimensional system of linear differential equations with constant coefficients

$$\frac{dX}{dt} = AX, \qquad X(0) = X_0, \tag{3.29}$$

where $\lambda_1 \ll \lambda_2 < 0$ are the eigenvalues of the matrix A, with respective eigenvectors a_1 and a_2 (of course, the computer does not know $\lambda_1, \lambda_2, a_1, a_2$).

We apply Euler's method to the problem (3.29):

$$X_{k+1} = X_k + AX_k h. \tag{3.30}$$

Let $X_0 = \alpha^1 a_1 + \alpha^2 a_2$. Then the solution of (3.29) has the form

$$X(t) = \alpha^1 e^{\lambda_1 t} a_1 + \alpha^2 e^{\lambda_2 t} a_2. \tag{3.31}$$

Since the first component of the solution decreases rapidly, because of the condition $\lambda_1 \ll \lambda_2 < 0$, on an initial interval of small length (equal to $\approx 1/|\lambda_1|$) we have to choose a very small step in (3.30). On the remainder of the interval (whose length we can compare in a natural way with $1/|\lambda_2|$), $X(t)$ changes slowly, and here the interpolation conditions do not require us to choose a small step h. We consider in more detail the nature of the method (3.30). We have

$$X_1 = (I + Ah)X_0 = (I + Ah)(\alpha^1 a_1 + \alpha^2 a_2) = (1 + \lambda_1 h)\alpha^1 a_1 + (1 + \lambda_2 h)\alpha^2 a_2.$$

It can be readily seen that

$$X_{k+1} = (1 + \lambda_1 h)^{k+1}\alpha^1 a_1 + (1 + \lambda_2 h)^{k+1}\alpha^2 a_2. \tag{3.32}$$

The computation by (3.32) will be suitable if h is chosen to satisfy $|1 + \lambda_1 h| \leq 1$, i.e.

$$h \leq \frac{2}{|\lambda_1|}. \tag{3.33}$$

Moreover, the condition (3.33) must be satisfied on the whole interval of length $\approx 1/|\lambda_2|$. Even when the first component in (3.31) has practically damped out, a step choice $h > 2/|\lambda_1|$ would, because of inevitable errors in the computations, again catch it, and a sharp increase in the error would result. Thus, when using the method (3.30) we have to choose the step very small (in accordance with (3.33)) on the whole interval of integration (of course, on the initial interval of length $\approx 1/|\lambda_1|$ the step must even be smaller, but this is because of natural causes and, in view of the smallness of $1/|\lambda_1|$, does not lead to any complications). The necessity of choosing the integration step small not only implies that the amount of computations increases, but also, more importantly, that the computational error increases. As a result, for appropriate λ_1, λ_2 this error may become so large that the method (3.30) becomes inapplicable for solving (3.29).

For $\lambda_1 \ll \lambda_2 < 0$ the system (3.29) belongs to the class of so-called *stiff systems* [39]. There is no unique generally accepted notion of stiffness, and different authors have proposed various definitions. Here it is better to talk about the phenomenon of stiffness, which is characterised, from the point of view of physics, by the presence of both fast as well as slow processes described by the system of differential equations. When solving such systems by explicit numerical integration methods there appears a mismatch between the necessity of choosing a very small integration step on the whole interval and the objective possibility of interpolating the solution on a large part of the interval with large step (since the solution changes slowly). Moreover, when using explicit methods, a small increase of the integration step within definite bounds leads to an explosion of the computational error.

Consider the implicit Euler method as applied to the system (3.29):

$$X_{k+1} = X_k + AX_{k+1}h. \tag{3.34}$$

We have

$$\begin{aligned} X_{k+1} &= (I - Ah)^{-1}X_k = (I - Ah)^{-(k+1)}X_0 \\ &= \frac{1}{(1-\lambda_1 h)^{k+1}}\alpha^1 a_1 + \frac{1}{(1-\lambda_2 h)^{k+1}}\alpha^2 a_2. \end{aligned} \tag{3.35}$$

It is clear from (3.35) that the method (3.34) does not have the property of instability, even not for arbitrary large h, i.e. when choosing h in (3.34) we need only worry about the error of the method. (It is clear that in the end the major differences in properties between the methods (3.30) and (3.34) are related with the various means of interpolating the exponentials $e^{\lambda_i t}$.) Of course, implicit methods are far more laborious than explicit methods, since in general they require one to solve at each step a system of nonlinear equations in X_{k+1}. To solve such systems a special method has been developed, [39], [43], with which we will not be concerned here.

A system of linear equations

$$\frac{dX}{dt} = AX + b(t), \tag{3.36}$$

involving a constant matrix A with eigenvalues λ_k, $k = 1, \dots, n$, having negative real parts, is called *stiff* (see [39], [43]) if the following condition holds:

$$\frac{\max_k |\mathrm{Re}\,\lambda_k|}{\min_k |\mathrm{Re}\,\lambda_k|} \gg 1. \tag{3.37}$$

A system of nonlinear equations is said to belong to the class of *stiff systems* if in a neighborhood of each point (t, x) in the domain under consideration its system of first approximation is stiff.

The quality, with respect to some measure, of a method is conveniently judged by the action of the method on some test system, having a small number of parameters and a simple form. To clarify stability properties of a method, one chooses as test system the equation

$$\frac{dX}{dt} = \lambda X, \tag{3.38}$$

where λ is a complex parameter with $\mathrm{Re}\,\lambda < 0$. The choice of (3.38) is related to the fact that every homogeneous system with constant coefficients having distinct eigenvalues with negative real parts can be decomposed into equations of the form (3.38).

The result of applying some method to (3.38) is a difference equation. For example, applying the explicit Euler method leads to the difference equation

$$X_{k+1} = X_k + \lambda h X_k, \tag{3.39}$$

while applying the implicit Euler method leads to

$$X_{k+1} = X_k + \lambda h X_{k+1}. \tag{3.40}$$

In similar difference equations λh enters as a parameter. For computational purposes it is natural to require that the trivial solution of such systems be stable, and to ensure applicability of the method for equations with arbitrary λ (requiring only $\mathrm{Re}\,\lambda < 0$) as h does not tend to zero we have to require stability for all λh belonging to the left halfplane of the complex λh-plane.

DEFINITION 3.1. The *region of stability* of a method is the set of values of λh satisfying the condition of asymptotic stability of the trivial solution of the difference equation arising when integrating the test equation (3.38) by this method. A method is called *A-stable* (absolutely stable) if the halfplane $\mathrm{Re}\,\lambda h < 0$ belongs to its region of stability.

There are other definitions of stability, but we will not be concerned with them. It is well known that no explicit Runge–Kutta or Adams method is A-stable. Hence, in particular, in relation with the numerical integration of stiff systems there arose the necessity of constructing implicit methods and of investigating their stability.

It can be readily seen from (3.40) that the implicit Euler method is A-stable, while (3.39) implies that the region of stability of the explicit Euler method is the interior of the disk with radius one and center at the point $\lambda h = -1$ in the complex λh-plane.

We now turn to the stochastic system with additive noises (3.1). It is natural to say that it is *stiff* if its deterministic part is stiff; as test equation we can naturally take

$$dX = \lambda X\, dt + \delta\, dw, \tag{3.41}$$

where λ is a complex parameter with $\operatorname{Re}\lambda < 0$ and σ is an arbitrary real parameter.

The method (3.24) as applied to equation (3.41) takes the form

$$\begin{aligned} X_{k+1} = {} & \left(1 + \alpha\lambda h + \beta(2\alpha - 1)\frac{(\lambda h)^2}{2}\right) X_k \\ & + \left((1-\alpha)\lambda h + (1-\beta)(2\alpha-1)\frac{(\lambda h)^2}{2}\right) X_{k+1} + \sigma\xi_k h^{1/2} \\ & + \lambda\sigma\left(\frac{2\alpha-1}{2}\xi_k + \frac{1}{\sqrt{12}}\eta_k\right) h^{3/2}. \end{aligned} \tag{3.42}$$

Equation (3.42) is a difference equation with additive noises. If for $\sigma = 0$ the trivial solution of (3.42) is asymptotically stable, then, in particular, for any σ any solution of (3.42) with $\mathbf{E}|X_0|^2 < \infty$ has second-order moments that are uniformly bounded in k. It is readily verified that in the opposite case the second-order moments tend to infinity as $k \to \infty$. Therefore the properties of the method (3.42) can be judged from the stability properties of (3.24) for $\sigma = 0$. Thus, e.g., to clarify the region of stability of a stochastic method we have to apply the method to equation (3.41) with $\sigma = 0$, i.e. to (3.38), and then clarify the stability properties of the difference equation obtained; the latter is clearly deterministic. In relation with this, Definition 3.1, concerning the region of stability and A-stability of a method, can be transferred without modifications to stochastic numerical integration methods.

EXAMPLE 3.2. Consider the method (3.28). Applying it to equation (3.38) gives the difference equation

$$X_{k+1} = \left(1 + \frac{\lambda h}{2}\right) X_k + \frac{\lambda h}{2} X_{k+1} < 1,$$

i.e.

$$X_{k+1} = \frac{1 + \frac{\lambda h}{2}}{1 - \frac{\lambda h}{2}} X_k.$$

It can be readily seen that if $\operatorname{Re}\lambda h < 0$, then

$$\left|\frac{1 + \frac{\lambda h}{2}}{1 - \frac{\lambda h}{2}}\right| < 1,$$

i.e. the region of stability includes the whole left halfplane of the complex λh-plane, and so the method (3.28) is A-stable.

EXAMPLE 3.3. Consider the method (3.24) for $\alpha = \beta = 0$. It can be readily computed that the region of stability of such a method is given by the inequality

$$\frac{1}{\left|1 - \lambda h + \frac{(\lambda h)^2}{2}\right|} < 1,$$

or, setting $\lambda h = \mu + i\nu$, by

$$\left|1 - (\mu + i\nu) + \frac{(\mu + i\nu)^2}{2}\right|^2 > 1.$$

The lefthand side of this inequality can be rewritten as follows:

$$\left(1 - \mu + \frac{\mu^2 - \nu^2}{2}\right)^2 + \nu^2(-1 + \mu)^2$$
$$= (1 - \mu)^2 + \left(\frac{\mu^2 - \nu^2}{2}\right)^2 + \mu^2(1 - \mu) + \nu^2\left((1 - \mu)^2 - (1 - \mu)\right),$$

which clearly exceeds 1 for $\mu < 0$. Hence the method under consideration is A-stable.

REMARK 3.1. The question of the stability of (implicit or explicit) methods in the case of systems with diffusion coefficients, depending on x, is far more complicated. In the class of linear autonomous stochastic systems

$$dX = AX\,dt + \sum_{r=1}^{q} B_r X\,dw_r(t) \tag{3.43}$$

we can regard as *stiff* those systems for which, first, the trivial solution is asymptotically stable, e.g. in mean-square, and, secondly, among the negative eigenvalues for the system of second-order moments for (3.43) there are eigenvalues with large as well as small modulus. We may hope that precisely such systems have, in a certain sense, both fast and slow processes. Here, as test system we can clearly take a second-order system of the form (3.43) with one-two noises and a small number of parameters. The application of a method to the test system leads to a stochastic linear difference system having a trivial solution. As a result there arises the possibility of judging the quality of the method by investigating the asymptotic and mean-square stability of this trivial solution.

All of the above represents not more than mere assumptions. It is clear that the establishment of the corresponding estimates in this direction requires the performance of a large amount of numerical experiments and of serious theoretical investigations.

3.4. Runge–Kutta type methods (implicit and explicit). In the method (3.11), the most complicated of all is the computation of $La(t_k, X_k)$. Using the idea of recomputation, we will construct a method in which $La(t_k, X_k)$ does not occur.

Introduce $\overline{X}^{(1)}(t+h) = \overline{X}(t+h)$ by Euler's method:

$$\overline{X}^{(1)}(t+h) = x + \sum_{r=1}^{q} \sigma_r(t)\xi_r h^{1/2} + a(t,x)h. \tag{3.44}$$

It can be readily seen that under the conditions of Theorem 3.1 we have for $\rho_1 = X_{t,x}(t+h) - \overline{X}^{(1)}(t+h)$ (recall that we are discussing a system with additive noises, for which Euler's method has order of accuracy one):

$$|\mathbf{E}\rho_1| \le K\left(1+|x|^2\right)^{1/2} h^2, \qquad (\mathbf{E}\rho_1)^{1/2} \le K\left(1+|x|^2\right)^{1/2} h^{3/2}. \tag{3.45}$$

Further,

$$a(t+h, X_{t,x}(t+h)) = a(t,x) + \sum_{r=1}^{q} \Lambda_r a(t,x)\xi_r h^{1/2} + La(t,x)h + \rho_2, \tag{3.46}$$

where

$$|\mathbf{E}\rho_2| \le K\left(1+|x|^2\right)^{1/2} h^2, \qquad (\mathbf{E}\rho_2)^{1/2} \le K\left(1+|x|^2\right)^{1/2} h. \tag{3.47}$$

Put $\rho_3 = a(t+h, X_{t,x}(t+h)) - a(t+h, \overline{X}^{(1)}(t+h))$. Since a satisfies a Lipschitz condition, by the second relation in (3.45) we successively have

$$\left(\mathbf{E}\rho_3^2\right)^{1/2} \le K\left(1+|x|^2\right)^{1/2} h^{3/2}, \qquad |\mathbf{E}\rho_3| \le \mathbf{E}|\rho_3| \le \left(\mathbf{E}\rho_3^2\right)^{1/2}. \tag{3.48}$$

We can write

$$La(t,x)h = a(t+h, \overline{X}^{(1)}(t+h)) - a(t,x) - \sum_{r=1}^{q} \Lambda_r a(t,x)\xi_r h^{1/2} + \rho_4, \tag{3.49}$$

where $\rho_4 = \rho_3 - \rho_2$.

By (3.47), (3.48) we have

$$|\mathbf{E}\rho_4| \le K\left(1+|x|^2\right)^{1/2} h^{3/2}, \qquad (\mathbf{E}\rho_4)^{1/2} \le K\left(1+|x|^2\right)^{1/2} h. \tag{3.50}$$

Consider the *one-step approximation* that is obtained from (3.11) by replacing in (3.11) $La(t,x)h$ (of course, in our context we also have to replace in (3.11) (t_k, X_k) by (t,x), ξ_{rk} by ξ_r, and η_{rk} by η_r) by the righthand side of (3.49) without the term ρ_4:

$$\begin{aligned}\overline{X}(t+h) = x &+ \sum_{r=1}^{q} \sigma_r(t)\xi_r h^{1/2} + \left(a(t,x) + a(t+h, \overline{X}^{(1)}(t+h))\right)\frac{h}{2} \\ &+ \sum_{r=1}^{q} \Lambda_r a(t,x)\frac{1}{\sqrt{12}}\eta_r h^{3/2} + \sum_{r=1}^{q} \frac{d\sigma_r}{dt}(t)\left(\frac{1}{2}\xi_r - \frac{1}{\sqrt{12}}\eta_r\right) h^{3/2}.\end{aligned} \tag{3.51}$$

The value $\overline{X}(t+h)$ computed by (3.51) differs from $\overline{X}(t+h)$ in the one-step approximation (3.11) by $\rho_4 h/2$. Therefore, for the $\overline{X}(t+h)$ in (3.51) we have

$$X(t+h) - \overline{X}(t+h) = \rho + \rho_4\frac{h}{2},$$

where ρ satisfies the relations (see the proof of (3.11) and the establishment of its order of accuracy)

$$\left|\mathbf{E}\rho^2\right| \le K\left(1+|x|^2\right)^{1/2} h^3, \qquad \left(\mathbf{E}\rho^2\right)^{1/2} \le K\left(1+|x|^2\right)^{1/2} h^2.$$

By (3.50) we have

$$\left|\mathbf{E}\left(\rho+\rho_4\frac{h}{2}\right)\right| = O(h^{5/2}), \qquad \left(\mathbf{E}\left(\rho+\rho_4\frac{h}{2}\right)^2\right)^{1/2} = O(h^2).$$

By Theorem 1.1, since $p_1 = 5/2$, $p_2 = 2$, the method corresponding to the one-step approximation (3.51) has order of accuracy 3/2. We state the result in a Theorem.

THEOREM 3.3. *Suppose that the coefficients $a(t,x)$ and $\sigma_r(t)$, $r = 1,\dots,q$, of equation* (3.1) *satisfy the conditions A. Then the method*

$$\begin{aligned} X_{k+1} = X_k &+ \sum_{r=1}^{q} \sigma_r(t_k)\xi_{rk}h^{1/2} \\ &+ \left(a(t_k,X_k) + \left(a\left(t_{k+1}, X_k + \sum_{r=1}^{q}\sigma_r(t_k)\xi_{rk}h^{1/2} + a(t_k,X_k)h\right)\right)\right)\frac{h}{2} \\ &+ \sum_{r=1}^{q}\Lambda_r a(t_k,X_k)\frac{1}{\sqrt{12}}\eta_{rk}h^{3/2} + \sum_{r=1}^{q}\frac{d\sigma_r}{dt}(t_k)\left(\frac{1}{2}\xi_{rk} - \frac{1}{\sqrt{12}}\eta_{rk}\right)h^{3/2}, \end{aligned} \tag{3.52}$$

where ξ_{rk} and η_{rk} are the same as in the method (3.11), *has order of accuracy* 3/2.

REMARK 3.2. In (3.51), $\Lambda_r a(t,x)$ ensures the existence of the first-order derivatives with respect to x and a (recall that $\Lambda_r a^i(t,x) = \sum_{j=1}^n \sigma_r^j(t)(\partial a^i/\partial x^j)(t,x))$. Therefore the method (3.52), which is based on (3.51), is an incomplete Runge–Kutta method. In fact, getting rid of the computation of the derivatives is not difficult. For example,

$$\frac{\partial a^i}{\partial x^j} \approx \frac{a^i(x^1,\dots,x^j+h,\dots,x^n) - a^i(x^1,\dots,x^j,\dots,x^n)}{h}.$$

Since this is an equality up to $O(h)$, by the assumptions made with respect to the function a^i, replacing in (3.52) all $\partial a^i/\partial x^j$ by their difference relations preserves the order of accuracy. However, this approach requires a large amount of recomputations. In the deterministic theory Runge–Kutta methods use a minimal amount of recomputations. Here we can also compute $\Lambda_r a(t,x)$ using only a single recomputation of the vector a. For this it suffices to use the identity

$$\Lambda_r a(t,x) = \frac{a(t,x+\sigma_r(t)h) - a(t,x)}{h} + O(h). \tag{3.53}$$

Note that $x + \sigma_r(t)h$ is the Euler approximation of the Cauchy problem

$$\frac{dY_t}{dt} = \sigma_r(t), \qquad Y_r(t) = x. \tag{3.54}$$

As a result, instead of (3.52) we can write down the following Runge–Kutta method:

$$
\begin{aligned}
X_{k+1}^{(1)} &= X_k + \sum_{r=1}^{q} \sigma_r(t_k)\xi_{rk}h^{1/2} + a(t_k, X_k)h, \\
Y_{k+1}^{(1)} &= X_k + \sigma_r(t_k)h, \\
X_{k+1} &= X_k + \sum_{r=1}^{q} \sigma_r(t_k)\xi_{rk}h^{1/2} + \left(a(t_k, X_k) + a(t_k, X_{k+1}^{(1)})\right)\frac{h}{2} \\
&+ \sum_{r=1}^{q} \left(a(t_k, Y_{k+1}^{(1)}) - a(t_k, X_k)\right)\frac{1}{\sqrt{12}}\eta_{rk}h^{1/2} \\
&+ \sum_{r=1}^{q} \frac{d\sigma_r}{dt}(t_k)\left(\frac{1}{2}\xi_{rk} - \frac{1}{\sqrt{12}}\eta_{rk}\right)h^{3/2}.
\end{aligned}
\tag{3.55}
$$

The idea of invoking, next to the initial system, other systems of differential equations, in the spirit of (3.53)–(3.54), in order to economise the amount of recomputations may turn out to be useful also in substantially more general situations. However, here we restrict ourselves to the remarks made above.

The method (3.52) is an explicit Runge–Kutta type method. We can construct implicit Runge–Kutta type methods by writing down the implicit versions of formulas (3.44) and (3.46), and preserve all remaining derivations. We can also substitute the righthand side of (3.49), without ρ_4, into (3.42) instead of into (3.11), putting $\beta = 1$ in (3.42). Having done the latter, for example, we obtain a one-parameter family of implicit Runge–Kutta type methods:

$$
\begin{aligned}
X_{k+1} &= X_k + \sum_{r=1}^{q} \sigma_r(t_k)\xi_{rk}h^{1/2} + a(t_k, X_k)\frac{h}{2} \\
&+ (1-\alpha)a(t_{k+1}, X_{k+1})h + \left(\alpha - \frac{1}{2}\right)a(t_{k+1}, \overline{X}^{(1)}(t_{k+1}))h \\
&+ \sum_{r=1}^{q} \Lambda_r a(t_k, X_k)\frac{1}{\sqrt{12}}\eta_{rk}h^{3/2} + \sum_{r=1}^{q} \sigma_r'(t_k)\left(\frac{1}{2}\xi_{rk} - \frac{1}{\sqrt{12}}\eta_{rk}\right)h^{3/2}.
\end{aligned}
\tag{3.56}
$$

Following the proofs of Theorem 3.2 and Theorem 3.3, it is not difficult to prove that the method (3.56) has order of accuracy 3/2 (here, the additional assumption that $La(t, x)$ has to satisfy a uniform Lipschitz condition in x can be dropped in the present situation).

3.5. Two-step difference methods. In (3.24) we put $\beta = 0$, $\alpha = \alpha_1 \neq 1/2$, and we express $La(t_{k+1}, X_{k+1})$ in terms of X_k, X_{k+1}, ξ_{rk}, and η_{rk}. Then we take $k+1$ instead of k in (3.24), put $\beta = 1$, $\alpha = \alpha_2$, and replace $La(t_{k+1}, X_{k+1})$ by the expression just found. As a result we obtain

$$\begin{aligned}
X_{k+1} &= \frac{1-2\alpha_2}{2\alpha_1-1}X_k + \frac{2(\alpha_1+\alpha_2-1)}{2\alpha_1-1}X_{k+1} + \frac{1-2\alpha_2}{2\alpha_1-1}\sum_{r=1}^{q}\sigma_r(t_k)\xi_{rk}h^{1/2} \\
&+ \sum_{r=1}^{q}\sigma_r(t_{k+1})\xi_{r(k+1)}h^{1/2} + \frac{\alpha_1(1-2\alpha_2)}{2\alpha_1-1}a(t_k,X_k)h \\
&+ \frac{4\alpha_1\alpha_2-\alpha_1-3\alpha_2+1}{2\alpha_1-1}a(t_{k+1},X_{k+1})h + (1-\alpha_2)a(t_{k+2},X_{k+2})h \\
&+ \frac{1-2\alpha_2}{2\alpha_1-1}\sum_{r=1}^{q}\Lambda_r a(t_k,X_k)\left(\frac{2\alpha_1-1}{2}\xi_{rk} + \frac{1}{\sqrt{12}}\eta_{rk}\right)h^{3/2} \\
&+ \frac{1-2\alpha_2}{2\alpha_1-1}\sum_{r=1}^{q}\frac{d\sigma_r}{dt}(t_k)\left(\frac{1}{2}\xi_{rk} - \frac{1}{\sqrt{12}}\eta_{rk}\right)h^{3/2} \\
&+ \sum_{r=1}^{q}\Lambda_r a(t_{k+1},X_{k+1})\left(\frac{2\alpha_2-1}{2}\xi_{r(k+1)} + \frac{1}{\sqrt{12}}\eta_{r(k+1)}\right)h^{3/2} \\
&+ \sum_{r=1}^{q}\frac{d\sigma_r}{dt}(t_{k+1})\left(\frac{1}{2}\xi_{r(k+1)} - \frac{1}{\sqrt{12}}\eta_{r(k+1)}\right)h^{3/2}.
\end{aligned} \tag{3.57}$$

For $\alpha_2 = 1/2$ the method (3.57) is the implicit one-step method (3.28) from Example 3.1 (with index k decreased by one in advance). For $\alpha_2 = 1$, $\alpha_1 \neq 1/2$ this is a one-parameter family of explicit two-step *difference methods*. For other α_2 and $\alpha_1 \neq 1/2$ this is a two-parameter family of implicit two-step difference methods. We cannot use Theorem 1.1 to study the order of accuracy of the method (3.57), since this Theorem is highly accomodated to one-step methods only.

THEOREM 3.4. *Suppose that, in addition to the conditions A, the function $\Lambda_r a$ satisfies a uniform Lipschitz condition in the variable x. Suppose*

$$0 \le \frac{1-2\alpha_2}{2\alpha_1-1} \le 1. \tag{3.58}$$

Then the method (3.57) *has order of accuracy equal to* 3/2 (*of course, under the assumptions that $X_0 = x(t_0)$, $X_1 = X(t_1)$*).

PROOF. Put $\beta_1 = (1-2\alpha_2)/(2\alpha_1-1)$, $\beta_2 = 2(\alpha_1+\alpha_2-1)/(2\alpha_1-1)$. Clearly, $\beta_1+\beta_2 = 1$, $0 \le \beta_1 \le 1$, $0 \le \beta_2 \le 1$. In (3.24) with $\beta = 0$, $\alpha = \alpha_1 \neq 1/2$, replace X_k and X_{k+1} by $X(t_k)$ and $X(t_{k+1})$. We obtain an equation P_1 (which we do not write out here). In distinction to (3.24), P_1 contains a righthand term: the remainder ρ_1. This remainder satisfies the inequalities (see the proof of (3.20)–(3.21))

$$|\mathbf{E}(\rho_1 \mid \mathcal{F}_{t_k})| \le K\left(1+|X(t_k)|^2\right)^{1/2}h^3, \qquad \left(\mathbf{E}\rho_1^2\right)^{1/2} \le K\left(1+\mathbf{E}|X(t_k)|^2\right)^{1/2}h^2. \tag{3.59}$$

From P_1 we can find $La(t_{k+1},X(t_{k+1}))h^2/2$. Now we take in (3.24) $k+1$ instead of k, and put $\beta = 1$, $\alpha = \alpha_2$ in it, so that (3.58) holds. Then we replace X_k, X_{k+1} by $X(t_k), X(t_{k+1})$, and we also replace $La(t_{k+1},X(t_{k+1}))h^2/2$ by the expression found for it from P_1. As a result we obtain an equation P. This equation differs from (3.57), first by the fact that we have $X(t_k), X(t_{k+1}), X(t_{k+2})$ instead of X_k, X_{k+1}, X_{k+2} in it, and secondly by the presence of a righthand term: the remainder

$$\rho_2 = \left(\frac{2\alpha_2 - 1}{2\alpha_1 - 1}\right)\rho_1 + \rho_2.$$

In other words, any exact solution satisfies (3.57) up to ρ; the remainder ρ satisfies precisely the relations (3.59), if only for a different constant K. Indeed, by construction ρ_2 satisfies relations of the form (3.59) with k replaced by $k+1$. Therefore,

$$\begin{aligned}
|\mathbf{E}(\rho_2 \mid \mathcal{F}_{t_k})| &= \left|\mathbf{E}\left(\mathbf{E}(\rho_2 \mid \mathcal{F}_{t_{k+1}}) \mid \mathcal{F}_{t_k}\right)\right| \\
&\le \mathbf{E}\left(\left|\mathbf{E}(\rho_2 \mid \mathcal{F}_{t_{k+1}})\right| \mid \mathcal{F}_{t_k}\right) \\
&\le \mathbf{E}\left(K\left(1 + |X(t_{k+1}|^2\right)^{1/2} h^3 \mid \mathcal{F}_{t_k}\right) \le K\left(1 + |X(t_k|^2\right)^{1/2} h^3
\end{aligned}$$

For ρ_2 the second inequality in (3.59) is even simpler to establish. Thus, the remainder ρ does satisfy the relations (3.59).

Subtract now termwise the equation (3.57) from P. We obtain:

$$\begin{aligned}
X(t_{k+2}) - X_{k+2} = {} & \beta_1 \left(X(t_k) - X_k\right) + \beta_2 \left(X(t_{k+1}) - X_{k+1}\right) \\
& + \gamma_1 \left(a(t_k, X(t_k)) - a(t_k, X_k)\right) h \\
& + \gamma_2 \left(a(t_{k+1}, X(t_{k+1})) - a(t_{k+1}, X_{k+1})\right) h \\
& + \gamma_3 \left(a(t_{k+2}, X(t_{k+2})) - a(t_{k+2}, X_{k+2})\right) h \\
& + \beta_1 \sum_{r=1}^{q} \left(\Lambda_r a(t_k, X(t_k)) - \Lambda_r a(t_k, X_k)\right) \mu_{rk} h^{3/2} \\
& + \sum_{r=1}^{q} \left(\Lambda_r a(t_{k+1}, X(t_{k+1})) - \Lambda_r a(t_{k+1}, X_{k+1})\right) \nu_{r(k+1)} h^{3/2} + \rho,
\end{aligned} \tag{3.60}$$

where $\gamma_1, \gamma_2, \gamma_3, \mu_{rk}, \nu_{r(k+1)}$ are notations for the corresponding constants and random variables. Square both sides of (3.60) and take the mathematical expectations of the expressions obtained. Putting $\epsilon_k^2 = \mathbf{E}|X(t_k) - X_k|^2$, we are led to the inequality

$$\begin{aligned}
\epsilon_{k+2}^2 \le {} & \beta_1^2 \epsilon_k^2 + 2\beta_1\beta_2 \epsilon_k \epsilon_{k+1} + \beta_2^2 \epsilon_{k+1}^2 + K\left(\epsilon_k^2 + \epsilon_{k+1}^2 + \epsilon_{k+2}^2\right) h \\
& + K\left(1 + \mathbf{E}|X(t_k)|^2\right)^{1/2} (\epsilon_k + \epsilon_{k+1}) h^2 + K\left(1 + \mathbf{E}|X(t_k)|^2\right) h^4.
\end{aligned} \tag{3.61}$$

In the proof of (3.61) we have used the Lipschitz property of a and $\Lambda_r a$, the Bunyakovsky–Schwarz inequality, and inequalities of the form

$$|\mathbf{E}\left(X(t_{k+1}) - X_{k+1}\right)\left(X(t_{k+2}) - X_{k+2}\right)| \le \epsilon_{k+1}\epsilon_{k+2} \le \frac{\epsilon_{k+1}^2 + \epsilon_{k+2}^2}{2},$$

$$\epsilon_k^2 h^3 \le \epsilon_k^2 h^2 \le \epsilon_k^2 h.$$

The expression $\epsilon_k h^2$ (and similarly $\epsilon_{k+1} h^2$) arose as upper bound for $\mathbf{E}\left(X(t_k) - X_k\right)\rho$, while Kh^4 is an upper bound for $\mathbf{E}\rho^2$.

We introduce

$$\epsilon'_0 = \epsilon_0 = 0, \qquad \epsilon'_1 = \epsilon_1 = 0,$$
$$\epsilon'_{k+2} = \beta_1^2 \epsilon_k'^2 + 2\beta_1\beta_2\epsilon'_k\epsilon'_{k+1} + \beta_2^2\epsilon_{k+1}'^2 + K\left(\epsilon_k'^2 + \epsilon_{k+1}'^2 + \epsilon_{k+2}'^2\right)h$$
$$+ K\left(1 + \mathbf{E}|X(t_k)|^2\right)^{1/2}\left(\epsilon'_k + \epsilon'_{k+1}\right)h^2 + K\left(1 + \mathbf{E}|X(t_k)|^2\right)h^4,$$
$$k = 0, \dots, N-2. \tag{3.62}$$

Clearly, $\epsilon'_k \geq \epsilon_k$, $\epsilon'_{k+1} \geq \epsilon'_k$. Replacing at the righthand side of (3.62) ϵ'_k by ϵ'_{k+1} and performing a number of elementary transformations (including, in particular, the use of the inequality $\epsilon'_{k+1}h^2 \leq \left(\epsilon_{k+1}'^2 h + h^3\right)/2)$, we are led to an inequality of the form

$$\epsilon_{k+2}'^2 \leq \epsilon_{k+1}'^2(1 + Kh) + K\left(1 + \mathbf{E}|X(t_0)|^2\right)h^3. \tag{3.63}$$

By Lemma 1.3, since $\epsilon'_1 = 0$ we hence obtain

$$\epsilon_k'^2 \leq K\left(1 + \mathbf{E}|X(t_0)|^2\right)h^2. \tag{3.64}$$

Consequently, also ϵ_k^2 satisfies the inequality (3.64).

We return once again to the inequality (3.60). Writing it for $X(t_{k+1}) - X_{k+1}$ in the form

$$X(t_{k+1}) - X_{k+1} = \beta_1\left(X(t_{k-1}) - X_{k-1}\right) + \beta_2\left(X(t_k) - X_k\right) + R,$$

we infer from it that

$$\mathbf{E}\left(X(t_{k+1} - X_{k+1}\right)\rho = \beta_1\mathbf{E}\left(X(t_{k-1}) - X_{k-1}\right)\rho + \beta_2\mathbf{E}\left(X(t_k) - X_k\right)\rho + \mathbf{E}R\rho. \tag{3.65}$$

Since ϵ_k^2 satisfies the inequality (3.64), it is easy to see that

$$\mathbf{E}R^2 \leq K\left(1 + \mathbf{E}|X_0|^2\right)h^4,$$

and consequently

$$\mathbf{E}|R\rho| \leq K\left(1 + \mathbf{E}|X_0|^2\right)h^4, \tag{3.66}$$

Further (see the first equality in (3.59), which is satisfied by ρ), for $j = k-1$ and $j = k$ we have

$$\begin{aligned} |\mathbf{E}\left((X(t_j) - X_j)\rho\right)| &= |\mathbf{E}\left((X(t_j) - X_j)\rho \mid \mathcal{F}_{t_k}\right)| \\ &= |\mathbf{E}\left((X(t_j) - X_j)\mathbf{E}(\rho \mid \mathcal{F}_{t_k})\right)| \\ &\leq \left(\mathbf{E}\left(X(t_j) - X_j\right)^2\right)^{1/2}\left(\mathbf{E}\left(\mathbf{E}(\rho \mid \mathcal{F}_{t_k})\right)^2\right)^{1/2} \\ &\leq K\left(1 + \mathbf{E}|X(t_0)|^2\right)h^4. \end{aligned} \tag{3.67}$$

From (3.65)–(3.67):

$$\mathbf{E}\left|\left(X(t_{k+1}) - X_{k+1}\right)\rho\right| \leq K\left(1 + \mathbf{E}|X(t_0)|^2\right)h^4. \tag{3.68}$$

By (3.67) for $j = k$ and (3.68), the inequality (3.61) can be refined using (3.60), in the sense that at the righthand side of (3.61) instead of $K\left(1 + \mathbf{E}|X(t_k)|^2\right)^{1/2}(\epsilon_k + \epsilon_{k+1})\, h^2$ there appears $K\left(1 + \mathbf{E}|X(t_0)|^2\right) h^4$. Reasoning as above, instead of (3.64) we obtain for ϵ_k^2:

$$\epsilon_k^2 \le K\left(1 + \mathbf{E}|X(t_0)|^2\right) h^3, \tag{3.69}$$

which proves Theorem 3.4. □

REMARK 3.3. The method (3.57) has the same features as difference methods in the deterministic situation. In it we do not compute La, while in comparison with the Runge–Kutta method it does not require recomputations. At the same time, to use it one has to look out for a value X_1 that is in practice sufficiently close to $X(t_1)$. As in the deterministic situation, to this end X_1 has to be found beforehand by using a one-step method that integrates the system (3.1) on the interval $[t_0, t_0 + h]$ with a small auxiliary step.

EXAMPLE 3.4. Consider the method (3.57) with $\alpha_1 = -1/2$, $\alpha_2 = 1$ (it is explicit). We investigate its A-stability. Applying it to the test equation (3.38) gives the difference equation

$$X_{k+2} = \left(\frac{1}{2} - \frac{1}{4}\lambda h\right) X_k + \left(\frac{1}{2} + \frac{7}{4}\lambda h\right) X_{k+1}.$$

It is easy to convince oneself that negative λh that are sufficiently large in absolute value do not belong to the region of stability. Therefore this method is not A-stable.

Consider now the method (3.57) with $\alpha_1 = 1$, $\alpha_2 = 0$. The corresponding difference equation has the form

$$X_{k+2} = (1 + \lambda h)X_k + \lambda h X_{k+2}.$$

Its trivial solution is asymptotically stable for all λh in the left halfplane. Therefore this method is A-stable.

4. Optimal integration methods for linear systems with additive noises

In this Section we will consider a linear system

$$dx = A(t)x\, dt + B(t)\, dw(t).$$

(Here we will use a somewhat different notation, which is more convenient in work on Kalman–Bucy filters.) Euler's method uses the increment $\Delta_m w = w(t_{m+1}) - w(t_m)$. It is natural to try to find a method that would optimally use this information regarding $w(t)$. As already noted in the Introduction, with respect to approximation in the mean-square sense the estimator

$$\hat{x}(t_k) = \mathbf{E}\left(x(t_k) \mid \Delta_m w,\ m = 0, 1, \ldots, k-1\right)$$

is best. If, next to $\Delta_m w$, additional information is known regarding the Wiener process, e.g., in the form of integrals $\int_{t_k}^{t_{k+1}} (\theta - t_k)\, dw(\theta)$, then the best approximation is given by the estimator

$$\hat{x}(t_k) = \mathbf{E}\left(x(t_k) \mid \Delta_m w,\ \int\limits_{t_k}^{t_{k+1}} (\theta - t)\, dw(\theta),\ m = 0, 1, \ldots, k-1\right).$$

The construction of these estimators rests on the solution of the optimal problem on numerical modeling of the Kalman–Bucy filter with discrete arrival of information. Therefore, in the first three Subsections below we give necessary results, from which simple recurrence methods for constructing the estimators follow as a simple consequence.

At the same time we have to note that for the system (1.4) we can construct exact integration methods if we use information regarding $w(t)$ in the form of integrals $\int_{t_i}^{t_{i+1}} F^{-1}(s)B(s)\, dw(s)$ (which have a Gaussian distribution), where $F(t)$ is the fundamental matrix of solutions of the system $dx/dt = A(t)x$. In this Section our aim is to show the restricted possibilities of having certain information regarding Wiener processes and to show the relation between numerical integration and optimal estimation.

4.1. Statement of the problem on numerical modeling of the Kalman–Bucy filter and on the optimal filter with discrete arrival of information. Consider the system

$$dx = A(t)x\, dt + B(t)\, dw(t), \tag{4.1}$$
$$dz = C(t)x\, dt + D(t)\, dv(t). \tag{4.2}$$

Here, x is an n-dimensional state vector, w is a k-dimensional perturbation vector, z is an m-dimensional observation vector, and v is an l-dimensional vector of observation errors. The matrices $A(t), B(t), C(t), D(t)$ have dimensions $n \times n$, $n \times k$, $m \times n$, and $m \times l$, respectively, and are continuous functions of time t in an interval $[t_0, T]$. The matrix $D(t)$ is such that the matrix $D(t)D^{\mathrm{T}}(t)$ is positive definite. The vectors w and v are standard Wiener processes, i.e. $\mathbf{E}dw = 0$, $\mathbf{E}dw\, dw^{\mathrm{T}} = I_{k\times k}\, dt$, $\mathbf{E}dv = 0$, $\mathbf{E}dv\, dv^{\mathrm{T}} = I_{l\times l}\, dt$, where $I_{p\times p}$ denotes the identity matrix of dimensions $p \times p$. We assume that, in general, $w(t)$ and $v(t)$ are mutually correlated:

$$\mathbf{E}dw\, dv^{\mathrm{T}} = A\, dt,$$

where S is a constant matrix of dimensions $k \times l$.

The initial state $x(t_0)$ has the Gaussian distribution with mathematical expectation $\mathbf{E}x(t_0) = m_0$ and covariance matrix $\mathbf{E}\,(x(t_0) - m_0)\,(x(t_0) - m_0)^{\mathrm{T}} = P_0$.

We assume that $x(t_0)$ does not depend on $w(t)$ or $v(t)$, $t \geq t_0$. The equations for the Kalman–Bucy filter for the system (4.1), (4.2) have the form (see, e.g., [19], [23],

[35]):

$$d\hat{x} = A(t)\hat{x}\,dt + K(t)\,(dz - C(t)\hat{x}\,dt)\,, \qquad \hat{x}(t_0) = m_0, \tag{4.3}$$

$$K(t) = \left(P(t)C^{\mathrm{T}}(t) + B(t)SD^{\mathrm{T}}(t)\right)\left(D(t)D^{\mathrm{T}}(t)\right)^{-1}, \tag{4.4}$$

$$\begin{aligned} \dot{P}(t) = &\left(A(t) - B(t)SD^{\mathrm{T}}(t)\left(D(t)D^{\mathrm{T}}(t)\right)^{-1}C(t)\right)P(t) \\ &+ P(t)\left(A(t) - B(t)SD^{\mathrm{T}}(t)\left(D(t)D^{\mathrm{T}}(t)\right)^{-1}C(t)\right)^{\mathrm{T}} \\ &- P(t)C^{\mathrm{T}}(t)\left(D(t)D^{\mathrm{T}}(t)\right)^{-1}C(t)P(t) \\ &- B(t)SD^{\mathrm{T}}(t)\left(D(t)D^{\mathrm{T}}(t)\right)^{-1}D(t)S^{\mathrm{T}}B^{\mathrm{T}}(t) + B(t)B^{\mathrm{T}}(t), \\ &P(t_0) = P_0. \end{aligned} \tag{4.5}$$

Here, $\hat{x}(t)$ denotes the optimal estimator for $x(t)$, and $P(t)$ is the covariance matrix of filtering errors:

$$P(t) = \mathbf{E}\,(x(t) - \hat{x}(t))\,(x(t) - \hat{x}(t))^{\mathrm{T}}\,. \tag{4.6}$$

In the numerical modeling of the *Kalman–Bucy filter*, when the incoming information $Z(t)$ is processed and used at discrete moments of time only, it is necessary to indicate a numerical integration method for the equation (4.3). For this one can use Euler's method:

$$\begin{aligned} \overline{x}_{i+1} = \overline{x}_i + A(t_i)\overline{x}_i h_i + K(t_i)\,(z(t_{i+1} - z(t_i) - C(t_i)\overline{x}_i h_i)\,, \\ \overline{x}_0 = m_0, \qquad i = 0, 1, \ldots, N-1, \end{aligned} \tag{4.7}$$

where $\overline{x}_i$ is an approximation of $\hat{x}(t_i)$; $t_0 < t_1 < \cdots < t_N = T$; and $h_i = t_{i+1} - t_i$.

The quantity $\overline{x}_i$ is an estimator for $x(t_i)$. Of course, this estimator is worse than $\hat{x}(t_i)$, which is related with the two following facts. First, in the construction of $\hat{x}(t_i)$ we use all information regarding $z(t)$ on the interval $[t_0, t_i]$, while at the same time $\overline{x}_i$ uses information regarding $z(t)$ at the discrete moments $t_0, t_1, \ldots, t_i$ only. Secondly, this discrete information is used by means of Euler's method (4.7) in a way that is not optimal. The best, in the mean-square sense, estimator $\hat{x}_i$ for $x(t_i)$ (to avoid ambiguity we note that $\hat{x}_i$ differs from $\hat{x}(t_i)$) with given observations $z(t_0), z(t_1), \ldots, z(t_i)$ is defined by the formula

$$\hat{x}_i = \mathbf{E}\,(x(t_i) \mid z(t_0), z(t_1), \ldots, z(t_i))\,. \tag{4.8}$$

Using a well-known property of the mathematical expectation we obtain

$$\begin{aligned} \hat{x}_i &= \mathbf{E}\,(x(t_i) \mid z(t_0), \ldots, z(t_i)) \\ &= \mathbf{E}\,(\mathbf{E}\,(x(t_i) \mid z(s),\ t_0 \le s \le t_i) \mid z(t_0), \ldots, z(t_i)) \\ &= \mathbf{E}\,(\hat{x}(t_i) \mid z(t_0), \ldots, z(t_i))\,. \end{aligned} \tag{4.9}$$

The equation (4.9) testifies of the fact that the optimal estimator $\hat{x}_i$ of the state $x(t_i)$ with discrete arrival of information is at the same time the best estimator for the solution $\hat{x}(t_i)$ of equation (4.3).

The problem of optimal estimation with discrete arrival of information was posed and solved in [20]. The basic construction in solving this problem is a special discretisation of the system (4.1), (4.2) (see [35], in which this discretisation is also considered). Here we will give the recurrence relations for $\hat{x}_i$ that have been obtained in [31]. Note that the characteristics of these relations differ in form from those given in [20], and are much more convenient for computational purposes.

4.2. Discretisation of the system (4.1), (4.2).

THEOREM 4.1. *The values of the state variables and of the observation output quantities of the system of stochastic differential equations* (4.1), (4.2) *at discrete moments of time t_i are related by the stochastic difference equations*

$$x(t_{i+1}) = F_i(t_{i+1})x(t_i) + w_i, \qquad x(t_0) = x_0, \tag{4.10}$$

$$z(t_{i+1}) = z(t_i) + C_i(t_{i+1})x(t_i) + v_i, \qquad z(t_0) = 0. \tag{4.11}$$

Here the matrices $F_i(t_{i+1})$ and $G_i(t_{i+1})$ can be found from the Cauchy problem for the system of matrix differential equations

$$\begin{aligned} \dot{F}_i &= A(t_i)F_i, \qquad F_i(t_i) = I_{n\times n}, \\ \dot{G}_i &= C(t)F_i, \qquad G_i(t_i) = O_{m\times n}, \end{aligned} \tag{4.12}$$

where $O_{m\times n}$ is the zero matrix of dimensions $m \times n$.

The processes w_i and v_i are n-dimensional and m-dimensional vector sequences of Gaussian variables with zero mean values and covariance matrices

$$\mathbf{E}w_iw_j^{\mathrm{T}} = \delta_{ij}Q_i(t_{i+1}), \qquad \mathbf{E}w_iv_j^{\mathrm{T}} = \delta_{ij}S_i(t_{i+1}), \qquad \mathbf{E}v_iv_j^{\mathrm{T}} = \delta_{ij}R_i(t_{i+1}), \tag{4.13}$$

where the matrices $Q_i(t_{i+1}), S_i(t_{i+1}), R_i(t_{i+1})$ can be found from the Cauchy problem for the system of matrix differential equations

$$\begin{aligned} \dot{Q}_i &= A(t)Q_i + Q_iA^{\mathrm{T}}(t) + B(t)B^{\mathrm{T}}(t), \qquad Q_i(t_i) = O_{n\times n}, \\ \dot{S}_i &= A(t)S_i + Q_iC^{\mathrm{T}}(t) + B(t)SD^{\mathrm{T}}(t), \qquad S_i(t_i) = O_{n\times m}, \\ \dot{R}_i &= C(t)S_i + S_i^{\mathrm{T}}C^{\mathrm{T}}(t) + D(t)D^{\mathrm{T}}(t), \qquad R_i(t_i) = O_{m\times m}. \end{aligned} \tag{4.14}$$

The initial state x_0 does not depend on the random sequences $\{w_i\}$ and $\{v_i\}$.

Theorem 4.1 differs from corresponding results in [20], [35] only in its determination of the characteristics of the system (4.10), (4.11). In [20] characteristics in the form of integrals are used; these are less convenient for a number of reasons, and their numerical realisation requires a larger amount of computational work than do (4.12)–(4.14). The relations (4.12)–(4.14) can be obtained from corresponding results in [20], [35].

For completeness of exposition we give a direct proof of these relations. Integrating (4.1) we obtain

$$x(t) = F_i(t)x(t_i) + \int_{t_i}^{t} F_i(t)F_i^{-1}(s)B(s)\,dw(s), \qquad t \geq t_i. \tag{4.15}$$

Substituting $x(t)$ into (4.2) we find

$$z(t) = z(t_i) + \int_{t_i}^{t} C(s)F_i(s)\,ds \cdot x(t_i) + \int_{t_i}^{t} C(s) \int_{t_i}^{s} F_i(s)F_i^{-1}(\tau)B(\tau)\,dw(\tau)\,ds + \int_{t_i}^{t} D(s)\,dv(s). \tag{4.16}$$

For $t = t_{i+1}$ the equations (4.15) and (4.16) become (4.10) and (4.11), if we put

$$w_i = \int_{t_i}^{t_{i+1}} F_i(t_{i+1})F_i^{-1}(s)B(s)\,dw(s),$$

$$v_i = \int_{t_i}^{t_{i+1}} C(s) \int_{t_i}^{s} F_i(s)F_i^{-1}(\tau)B(\tau)\,dw(\tau)\,ds + \int_{t_i}^{t_{i+1}} D(s)\,dv(s).$$

To compute the covariance matrices $Q_i(t_{i+1}), S_i(t_{i+1}), R_i(t_{i+1})$ we consider the system of stochastic differential equations

$$\begin{aligned} d\eta_i &= A(t)\eta_i\,dt + B(t)\,dw(t), \qquad \eta_i(t_i) = 0,\\ d\zeta_i &= C(t)\eta_i\,dt + D(t)\,dv(t), \qquad \zeta_i(t_i) = 0. \end{aligned} \tag{4.17}$$

It can be readily seen that $w_i = \eta_i(t_{i+1})$, $v_i = \zeta_i(t_{i+1})$. Applying Itô's formula, we go from (4.17) to the system

$$\begin{aligned} d(\eta_i\eta_i^{\mathrm{T}}) &= (A(t)\eta_i\,dt + B(t)\,dw(t))\,\eta_i^{\mathrm{T}}\\ &\quad + \eta_i\,(A(t)\eta_i\,dt + B(t)\,dw(t))^{\mathrm{T}} + B(t)B^{\mathrm{T}}(t)\,dt,\\ d(\eta_i\zeta_i^{\mathrm{T}}) &= (A(t)\eta_i\,dt + B(t)\,dw(t))\,\zeta_i^{\mathrm{T}}\\ &\quad + \eta_i\,(C(t)\eta_i\,dt + D(t)\,dv(t))^{\mathrm{T}} + B(t)SD^{\mathrm{T}}(t)\,dt,\\ d(\zeta_i\zeta_i^{\mathrm{T}}) &= (C(t)\eta_i\,dt + D(t)\,dv(t))\,\zeta_i^{\mathrm{T}}\\ &\quad + \zeta_i\,(C(t)\eta_i\,dt + D(t)\,dv(t))^{\mathrm{T}} + D(t)D^{\mathrm{T}}(t)\,dt, \end{aligned} \tag{4.18}$$

Taking the mathematical expectation of both sides of the equations in (4.18) we obtain the system (4.14). This proves Theorem 4.1.

4.3. An optimal filter with discrete arrival of information. The best estimator $\hat{x}_i$ for $x(t_i)$ given observations $z(t_0), \dots, z(t_i)$ is determined by (4.8). By Theorem 4.1, the variables $x(t_i)$ and $z(t_i)$ satisfy the linear system of stochastic difference equations (4.10)–(4.11). By constructing the *Kalman–Bucy filter* for the discrete system (4.10)–(4.11) we obtain recurrence relations for $\hat{x}_i$. We write out the equations for the Kalman–Bucy filter for the system (4.10)–(4.11) [23]:

$$\hat{x}_{i+1} = F_i(t_{i+1})\hat{x}_i + K_i\left(z(t_{i+1}) - z(t_i) - G_i(t_{i+1})\hat{x}_i\right), \qquad \hat{x}_0 = m_0, \tag{4.19}$$

$$K_i = \left(F_i(t_{i+1})P_iG_i^{\mathrm{T}}(t_{i+1}) + S_i(t_{i+1})\right)\left(G_i(t_{i+1})P_iG_i^{\mathrm{T}}(t_{i+1}) + R_i(t_{i+1})\right)^{-1}, \tag{4.20}$$

$$P_{i+1} = F_i(t_{i+1})P_iF_i^{\mathrm{T}}(t_{i+1}) + Q_i(t_{i+1}) - K_i\left(F_i(t_{i+1})P_iG_i^{\mathrm{T}}(t_{i+1}) + S_i(t_{i+1})\right)^{\mathrm{T}}, \tag{4.21}$$

$$i = 0, 1, \ldots, N-1,$$

where P_i is the covariance matrix of filtering errors:

$$P_i = \mathbf{E}\left(x(t_i) - \hat{x}_i\right)\left(x(t_i) - \hat{x}_i\right)^{\mathrm{T}}.$$

We sum up this result in the following Theorem.

THEOREM 4.2. *The optimal estimator* $\hat{x}_i = \mathbf{E}\left(x(t_i) \mid z(t_0), \ldots, z(t_i)\right)$ *of the state* $x(t_i)$ *for the system* (4.1)–(4.2) *with discrete arrival of information is determined by the recurrence relations* (4.19)–(4.21). *These same variables* $\hat{x}_i$ *give the best, in mean-square sense, approximate solution of the filtering equations* (4.3) *at the moment* t_i.

REMARK 4.1. Here we have derived the optimal estimator for systems with non-singular noises in the observations. Indeed, this suffices for the aims investigated by us. However, all reasonings can be transferred, without serious changes, to systems with singular noises in the observations. At certain places we have to take pseudo-inverse matrices instead of inverse matrices. We have to note that for systems with singular noises in the observations, which are often met in practice, there is in general no acceptable formula for $\hat{x}(t) = \mathbf{E}(x(t) \mid z(s),\ t_0 \le s \le t)$, i.e. there is no concrete method for constructing the continuous Kalman–Bucy filter. The method of discrete approximation, which allows one to constructively obtain in the most general case an approximate value for $\hat{x}(t_i)$, is at the same time a regularisation method for the continuous Kalman–Bucy filter for systems with singular noises in the observations (see [32]).

4.4. An optimal integration method of the first order of accuracy. We use the above results for constructing a numerical integration method for the system (4.1) which uses in an optimal manner the discrete process Δw_i being modeled. The best mean-square approximation of $x(t_i)$ is given by the estimator

$$\hat{x}_i = \mathbf{E}\left(x(t_i) \mid \Delta w_0, \ldots, \Delta w_{i-1}\right). \tag{4.22}$$

If we introduce the auxiliary system of stochastic differential equations

$$dz = dw, \qquad z(t_0) = 0, \tag{4.23}$$

we can write

$$\hat{x}_i = \mathbf{E}\left(x(t_i) \mid \Delta w_0, \ldots, \Delta w_{i-1}\right) = \mathbf{E}\left(x(t_i) \mid z(t_1), \ldots, z(t_i)\right). \tag{4.24}$$

Now we use Theorem 4.2. Applying it to the system (4.1), (4.23) gives for $\hat{x}_i$ the recurrence relation

$$\hat{x}_{i+1} = F_i(t_{i+1})\hat{x}_i + \frac{1}{h_i} S_i(t_{i+1})\Delta w_i, \qquad \hat{x}_0 = x_0, \tag{4.25}$$

where the matrix $F_i(t_{i+1})$ can be found from (4.12), and $S_i(t_{i+1})$ can be found from the Cauchy problem for the matrix differential equation

$$\dot{S}_i = A(t)S_i + B(t), \qquad S_i(t_i) = O_{n\times k}. \tag{4.26}$$

We also give a recurrence relation for the covariance matrix $P_i = \mathbf{E}\,(x(t_i) - \hat{x}_i)\,(x(t_i) - \hat{x}_i)^{\mathrm{T}}$:

$$P_{i+1} = F_i(t_{i+1})P_iF_i^{\mathrm{T}}(t_{i+1}) + Q_i(t_{i+1}) - \frac{1}{h_i} S_iS_i^{\mathrm{T}}(t_{i+1}), \qquad P_0 = O_{n\times n}, \tag{4.27}$$

where $Q_i(t_{i+1})$ can be found from (4.14).

Note that P_i does not occur in (4.25). It is clear that since the method (4.25) is optimal, it has order of accuracy not less than that of Euler's method, i.e. not less than $O(h)$ (recall that for systems with additive noises Euler's method has order of accuracy equal to one). We show that this order is exactly $O(h)$, as for Euler's method. To this end we consider the scalar equation

$$dx = ax\,dt + b\,dw(t), \qquad x(t_0) = x_0, \tag{4.28}$$

with constant coefficients $a \neq 0$, $b \neq 0$. For $p_i = \mathbf{E}\,(x(t_i) - \hat{x}_i)^2$ we obtain from (4.27) the recurrence equation

$$p_{i+1} = e^{2ah}p_i + \frac{1}{12}a^2b^2h^3 + O(h^4), \qquad p_0 = 0.$$

This gives

$$p_N = \left(\frac{1}{12}a^2b^2h^3 + O(h^4)\right)\frac{e^{2a(T-t_0)} - 1}{e^{2ah} - 1} = O(h^2).$$

Thus, there is no numerical integration formula for the system (4.1) that uses only information about $w(t)$ at discrete moments of time t_i, $i = 0, \dots, N$, and that would have order of accuracy higher than $O(h)$ (see also [47]). Note that although the method (4.25) has the same order of accuracy as Euler's method, in a number of cases it gives far more precise results.

REMARK 4.2. We show that the method (4.25) is A-stable. Indeed, applying it to the test equation

$$dx = \lambda x\,dt$$

gives the difference equation

$$\hat{x}_{i+1} = e^{\lambda h}\hat{x}_i,$$

and $|e^{\lambda h}| < 1$ for $\operatorname{Re}\lambda h < 0$.

In fact, the method (4.25) requires in its construction the values $F_i(t_{i+1})$ and $S_i(t_{i+1})$, which comes down to exactly solving the problems (4.12) and (4.26). For a stiff system (4.1) the sufficiently exact numerical solution of these problems meets with certain difficulties and, in the end, requires a special approach (see, e.g., [39]).

REMARK 4.3. It is clear that the construction of an optimal integration method in the case that, next to Δw_i, also the $\int_{t_i}^{t_{i+1}}(\theta - t_i)\, dw(\theta)$ are known reduces to applying Theorem 4.2 to the system consisting of (4.1) and the equation

$$dz^{(1)} = dw, \qquad dz^{(2)} = t\, dw.$$

It is obvious that the order of accuracy of this method is at least 3/2, since by these means for any system with additive noises we can construct a method of order 3/2. We can prove that the order of this method is precisely 3/2.

5. A strengthening of the main convergence theorem

The aim of this Section is to obtain an estimate of the form (1.8) for

$$\left[\mathbf{E} \max_k \left| X_{t_0,X_0}(t_k) - \overline{X}_{t_0,X_0}(t_k)\right|^2\right]^{1/2}.$$

Let η_k denote the deviation $X_{t_0,X_0}(t_k) - \overline{X}_{t_0,X_0}(t_k)$. For the time being we assume that $(\eta_k^2, \mathcal{F}_{t_k})$, $k = 0, \dots, N$, is a submartingale, and that $\mathbf{E}\eta_N^4 < \infty$. Then (see [23, p. 49]):

$$\mathbf{E}\left(\max_k \eta_k^2\right) \le \left[\mathbf{E}\left(\max_k \eta_k^4\right)\right]^{1/2} \le \left(4\mathbf{E}\eta_N^4\right)^{1/2}. \tag{5.1}$$

If also $\mathbf{E}\eta_N^4 = O(h^{4p_2-2})$, then (5.1) implies the required estimate. Although the sequence η_k^2 is not itself a submartingale, it is close to being submartingale; in the sequel we will prove and use this correspondingly. The relation $\mathbf{E}\eta_N^4 = O(h^{4p_2-2})$ is sufficiently natural, since $\mathbf{E}\eta_N^2 = O(h^{2p_2-1})$. In Theorem 1.1 the latter relation follows from the fact that the mean of the square of the one-step error is $O(h^{2p_2})$. Therefore we may expect that $\mathbf{E}\eta_N^4 = O(h^{4p_2-2})$ follows from the fact that the mean of the fourth power of the one-step error is only $O(h^{4p_2-1})$. In Theorem 5.1 which, next to being of auxiliary interest, is also of interest in its own right, this weak assumption is made (as a rule, the mean of the fourth power will be $O(^{4p_2})$, since the mean of the second power is $O(h^{2p_2})$ by (1.6)).

5.1. The theorem on convergence in the mean of order 4.

THEOREM 5.1. *In addition to the conditions of Theorem* 1.1, *suppose that also*

$$\left[\mathbf{E}\left|X_{t,x}(t+h) - \overline{X}_{t,x}(t+h)\right|^4\right]^{1/4} \le K\left(1+|x|^4\right)^{1/4} h^{p_2-1/4}, \tag{5.2}$$

and let $p_2 \ge 3/4$. *Then for arbitrary* N *and* $k = 0, \dots, N$,

$$\left(\mathbf{E}\eta_k^4\right)^{1/4} = \left(\mathbf{E}\left|X_{t_0,x_0}(t_k) - \overline{X}_{t_0,x_0}(t_k)\right|^4\right)^{1/4} \le K\left(1+\mathbf{E}|X_0|^4\right)^{1/4} h^{p_2-1/2}. \tag{5.3}$$

To a certain extent the proof of this Theorem repeats the proof of Theorem 1.1. Therefore we do not give it in as much detail as we have given the proof of Theorem 1.1.

Without further mentioning it, everywhere below we assume that $\mathbf{E}|X_0|^4 < \infty$.

LEMMA 5.1. *There is a representation*

$$X_{t,x}(t+h) - X_{t,y}(t+h) = x - y + Z \tag{5.4}$$

for which

$$\mathbf{E}\,|X_{t,x}(t+h) - X_{t,y}(t+h)|^4 \le |x-y|^4(1+Kh), \tag{5.5}$$

$$\mathbf{E}Z^4 \le K\,|x-y|^4h^2. \tag{5.6}$$

We defer the proof of this Lemma until after the following Remark, which concerns notations.

REMARK 5.1. Let α and β be vectors of the same dimension. By $\alpha\beta = \beta\alpha$ we denote the scalar product of these vectors. In particular, α^2 denotes the scalar product of α with itself, and so $|\alpha|^2 = \alpha^2$. Further, $\alpha^3\beta$ is the product of the two scalars α^2 and $\alpha\beta$. The notation without parentheses is possible because $\alpha^2(\alpha\beta) = \alpha(\alpha^2)\beta = (\beta\alpha)\alpha^2$ etc. At the same time $(\alpha\beta)^2$ is not equal to $\alpha^2\beta^2$, as a rule. We also note the following inequalities $(\alpha+\beta)^2 = \alpha^2 + 2\alpha\beta + \beta^2 \le 2\alpha^2 + 2\beta^2$, $(\alpha+\beta)^4 \le 8\alpha^4 + 8\beta^4$. In the sequel we will use similar notations without additional clarifications.

PROOF OF LEMMA 5.1. Introduce the scalar

$$\mathcal{U}(\theta) = (X_{t,x}(\theta) - X_{t,y}(\theta))^2.$$

By Itô's formula we have

$$\begin{aligned}
d\mathcal{U} &= 2\,(X_{t,x}(\theta) - X_{t,y}(\theta))\,(a(\theta, X_{t,x}(\theta)) - a(\theta, X_{t,y}(\theta)))\,d\theta \\
&\quad + \sum_{r=1}^{q} (\sigma_r(\theta, X_{t,x}(\theta)) - \sigma_r(\theta, X_{t,y}(\theta)))^2\,d\theta \\
&\quad + 2\,(X_{t,x}(\theta) - X_{t,y}(\theta)) \sum_{r=1}^{q} (\sigma_r(\theta, X_{t,x}(\theta)) - \sigma_r(\theta, X_{t,y}(\theta)))\,dw_r(\theta).
\end{aligned}$$

Again by Itô's formula we obtain

$$\begin{aligned}
d\,&(X_{t,x}(\theta) - X_{t,y}(\theta))^4 = d\mathcal{U}^2 = 2\mathcal{U}\,d\mathcal{U} + (d\mathcal{U})^2 \\
&= 4\,(X_{t,x}(\theta) - X_{t,y}(\theta))^2 \\
&\quad \times [(X_{t,x}(\theta) - X_{t,y}(\theta))\,(a(\theta, X_{t,x}(\theta)) - a(\theta, X_{t,y}(\theta)))\,d\theta] \\
&\quad + 2\,(X_{t,x}(\theta) - X_{t,y}(\theta))^2 \sum_{r=1}^{q} (\sigma_r(\theta, X_{t,x}(\theta)) - \sigma_r(\theta, X_{t,y}(\theta)))^2\,d\theta \\
&\quad + 4 \sum_{r=1}^{q} (X_{t,x}(\theta) - X_{t,y}(\theta))\,(\sigma_r(\theta, X_{t,x}(\theta)) - \sigma_r(\theta, X_{t,y}(\theta)))^2\,d\theta \\
&\quad + 2\,(X_{t,x}(\theta) - X_{t,y}(\theta))^2 \\
&\quad \times \left[(X_{t,x}(\theta) - X_{t,y}(\theta)) \sum_{r=1}^{q} (\sigma_r(\theta, X_{t,x}(\theta)) - \sigma_r(\theta, X_{t,y}(\theta)))\,dw_r(\theta)\right].
\end{aligned}$$

To arrive at the inequality (5.5) we have to integrate this equation from t to $t+h$, take the mathematical expectation, and use the Lipschitz and Gronwall inequalities.

We turn to the proof of (5.6). We have:

$$\mathbf{E}Z^4 \le 8\mathbf{E}\left(\int_t^{t+h}\sum_{r=1}^q (\sigma_r(s,X_{t,x}(s)) - \sigma_r(s,X_{t,y}(s)))\,dw_r(s)\right)^4 + 8\mathbf{E}\left(\int_t^{t+h}(a(s,X_{t,x}(s) - X_{t,y}(s)))\,ds\right)^4.$$

The first integral can be estimated, using the inequality from [8, p. 27], the Lipschitz inequality, and the inequality (5.5) already obtained, as follows:

$$\mathbf{E}\left(\int_t^{t+h}\sum_{r=1}^q (\sigma_r(s,X_{t,x}(s)) - \sigma_r(s,X_{t,y}(s)))\,dw_r(s)\right)^4 \le Kh\int_t^{t+h}\mathbf{E}\left(\sigma_r(s,X_{t,x}(s)) - \sigma_r(s,X_{t,y}(s))\right)^4 ds \le K\,|x-y|^4h^2.$$

It is easy to see that the second integral is $O(h^4)\cdot(x-y)^4$. □

LEMMA 5.2. *Let $p_2 \ge 3/4$. Then for all natural numbers N and all $k = 0,\dots,N$ we have*

$$\mathbf{E}\left|\overline{X}_k\right|^4 \le K\left(1 + \mathbf{E}|X_0|^4\right). \tag{5.7}$$

PROOF. The existence of $\mathbf{E}\left|\overline{X}_k\right|^4$ for all $k = 0,\dots,N$ can be proved similarly as in Lemma 1.2. Write

$$\overline{X}_{k+1} = \overline{X}_k + \left(\overline{X}_{k+1} - \overline{X}_k\right),$$

$$\overline{X}_{k+1}^2 = \overline{X}_k^2 + \left(\overline{X}_{k+1} - \overline{X}_k\right)^2 + 2\overline{X}_k\left(\overline{X}_{k+1} - \overline{X}_k\right),$$

$$\begin{aligned}\overline{X}_{k+1}^4 = \overline{X}_k^4 &+ \left(\overline{X}_{k+1} - \overline{X}_k\right)^4 + 4\left(\overline{X}_k\left(\overline{X}_{k+1} - \overline{X}_k\right)\right)^2 + 2\overline{X}_k^2\left(\overline{X}_{k+1} - \overline{X}_k\right)^2 \\ &+ 4\overline{X}_k^3\left(\overline{X}_{k+1} - \overline{X}_k\right) + 4\overline{X}_k\left(\overline{X}_{k+1} - \overline{X}_k\right)^3.\end{aligned} \tag{5.8}$$

We have

$$\left|\overline{X}_{k+1} - \overline{X}_k\right| \le \left|X_{t_k,\overline{X}_k}(t_{k+1}) - \overline{X}_k\right| + \left|X_{t_k,\overline{X}_k}(t_{k+1}) - \overline{X}_{t_k,\overline{X}_k}(t_{k+1})\right|,$$

$$\left|\overline{X}_{k+1} - \overline{X}_k\right|^4 \le 8\left|X_{t_k,\overline{X}_k}(t_{k+1}) - \overline{X}_k\right|^4 + 8\left|X_{t_k,\overline{X}_k}(t_{k+1}) - \overline{X}_{t_k,\overline{X}_k}(t_{k+1})\right|^4. \tag{5.9}$$

Since (see [8, p. 48])

$$\mathbf{E}\left|X_{t_k,\overline{X}_k}(t_{k+1}) - \overline{X}_k\right|^4 \le K\left(1 + \mathbf{E}\left|\overline{X}_k\right|^4\right)h^2 \tag{5.10}$$

and, by the conditional version of (5.2),

$$\mathbf{E}\left|X_{t_k,\overline{X}_k}(t_{k+1}) - \overline{X}_{t_k,\overline{X}_k}(t_{k+1})\right|^4 \leq K\left(1+\mathbf{E}\left|\overline{X}_k\right|^4\right) h^{4p_2-1}, \tag{5.11}$$

we have (recall that $p_2 \geq 3/4$)

$$\mathbf{E}\left(\overline{X}_{k+1} - \overline{X}_k\right)^4 \leq K\left(1+\mathbf{E}\left|\overline{X}_k\right|^4\right) h^2. \tag{5.12}$$

Further, $\left(\overline{X}_k\left(\overline{X}_{k+1}-\overline{X}_k\right)\right)^2 \leq \overline{X}_k^2\left(\overline{X}_{k+1}-\overline{X}_k\right)^2$ and, by (5.12),

$$\mathbf{E}\left(\overline{X}_k^2\left(\overline{X}_{k+1}-\overline{X}_k\right)^2\right) \leq \left(\mathbf{E}\overline{X}_k^4\right)^{1/2}\left(\mathbf{E}\left(\overline{X}_{k+1}-\overline{X}_k\right)^4\right)^{1/2} \leq K\left(1+\mathbf{E}\left|\overline{X}_k\right|^4\right) h. \tag{5.13}$$

Using the Hölder inequality we find

$$\left|\mathbf{E}\overline{X}_k\left(\overline{X}_{k+1}-\overline{X}_k\right)^3\right| \leq \left(\mathbf{E}\overline{X}_k^4\right)^{1/4}\left(\mathbf{E}\left|\overline{X}_{k+1}-\overline{X}_k\right|^{3\cdot4/3}\right)^{3/4} \leq K\left(1+\mathbf{E}\overline{X}_k^4\right) h^{6/4}. \tag{5.14}$$

Finally we will estimate

$$\left|\mathbf{E}\overline{X}_k^3\left(\overline{X}_{k+1}-\overline{X}_k\right)\right| = \left|\mathbf{E}\overline{X}_k^3\mathbf{E}\left(\left(\overline{X}_{k+1}-\overline{X}_k\right) \mid \mathcal{F}_{t_k}\right)\right|.$$

To this end we write

$$\begin{aligned}\mathbf{E}\left(\left(\overline{X}_{k+1}-\overline{X}_k\right) \mid \mathcal{F}_{t_k}\right) &= \mathbf{E}\left(\left(X_{t_k,\overline{X}_k}(t_{k+1})-\overline{X}_k\right) \mid \mathcal{F}_{t_k}\right)\\ &\quad + \mathbf{E}\left(\left(\overline{X}_{t_k,\overline{X}_k}(t_{k+1}) - X_{t_k,\overline{X}_k}(t_{k+1})\right) \mid \mathcal{F}_{t_k}\right).\end{aligned}$$

The modulus of the first term is bounded by $K\left(1+\overline{X}_k^2\right)^{1/2} h$, and, by (1.5), the modulus of the second term is bounded by $K\left(1+\overline{X}_k^2\right)^{1/2} h^{p_1}$. Whence, since $p_1 \geq p_2 + 1/2 > 1$, it is easy to find that

$$\mathbf{E}\left|\mathbf{E}\left(\left(\overline{X}_{k+1}-\overline{X}_k\right) \mid \mathcal{F}_{t_k}\right)\right|^4 \leq K\left(1+\mathbf{E}\left|\overline{X}_k\right|^4\right) h^4.$$

Therefore Hölder's inequality implies

$$\begin{aligned}\left|\mathbf{E}\overline{X}_k^3\left(\overline{X}_{k+1}-\overline{X}_k\right)\right| &\leq \left(\mathbf{E}\left|\overline{X}_k\right|^{3\cdot4/3}\right)^{3/4}\left(\mathbf{E}\left|\mathbf{E}\left(\left(\overline{X}_{k+1}-\overline{X}_k\right) \mid \mathcal{F}_{t_k}\right)\right|^4\right)^{1/4}\\ &\leq K\left(1+\mathbf{E}\left|\overline{X}_k\right|^4\right) h.\end{aligned} \tag{5.15}$$

Using (5.12)–(5.15), equation (5.8) implies

$$\mathbf{E}\overline{X}_{k+1}^4 \leq \mathbf{E}\overline{X}_k^4 + K\left(1+\mathbf{E}\left|\overline{X}_k\right|^4\right) h.$$

By Lemma 1.3 this implies (5.7). □

We introduce the following notations:

$$\eta_{k+1} = X_{t_0,X_0}(t_{k+1}) - \overline{X}_{t_0,X_0}(t_{k+1}) = X_{t_k,X(t_k)}(t_{k+1}) - \overline{X}_{t_k,\overline{X}(t_k)}(t_{k+1}) = \alpha_{k+1} + \rho_{k+1},$$
$$\alpha_{k+1} = X_{t_k,X(t_k)}(t_{k+1}) - X_{t_k,\overline{X}(t_k)}(t_{k+1}),$$
$$\rho_{k+1} = X_{t_k,\overline{X}(t_k)}(t_{k+1}) - \overline{X}_{t_k,\overline{X}(t_k)}(t_{k+1}).$$

We have

$$\eta_{k+1}^4 = \alpha_{k+1}^4 + 4\alpha_{k+1}^3\rho_{k+1} + a(\alpha_{k+1}\rho_{k+1})^2 + 4\alpha_{k+1}\rho_{k+1}^3 + \rho_{k+1}^4 + 2\alpha_{k+1}^2\rho_{k+1}^2. \tag{5.16}$$

We write

$$\alpha_{k+1} = X(t_k) - \overline{X}_k + z_{k+1} = \eta_k + z_{k+1}. \tag{5.17}$$

By the conditional version of Lemma 5.1 we have

$$\mathbf{E}\alpha_{k+1}^4 \le \mathbf{E}\eta_k^4(1+Kh), \tag{5.18}$$
$$\mathbf{E}\left(z_{k+1} \mid \mathcal{F}_{t_k}\right) \le K\eta_k^4 h^2. \tag{5.19}$$

We introduce the notation $\delta_k = \left(\mathbf{E}\eta_k^4\right)^{1/4}$.

LEMMA 5.3.

$$\left|\mathbf{E}\left(\alpha_{k+1}^3\rho_{k+1}\right)\right| \le K\left(1+\mathbf{E}\left|X_0\right|^4\right)^{1/4}\delta_k^3 h^{p_2+1/2}. \tag{5.20}$$

PROOF. We first estimate

$$\begin{aligned}\mathbf{E}\left(\alpha_{k+1}^3\rho_{k+1} \mid \mathcal{F}_{t_k}\right) &= \mathbf{E}\left(\eta_k^3\rho_{k+1} \mid \mathcal{F}_{t_k}\right) + \mathbf{E}\left(\eta_k^2 z_{k+1}\rho_{k+1} \mid \mathcal{F}_{t_k}\right)\\ &\quad + 2\mathbf{E}\left(\eta_k z_{k+1}\cdot\eta_k\rho_{k+1} \mid \mathcal{F}_{t_k}\right) + 2\mathbf{E}\left(\eta_k z_{k+1}\cdot z_{k+1}\rho_{k+1} \mid \mathcal{F}_{t_k}\right)\\ &\quad + \mathbf{E}\left(z_{k+1}^2\eta_k\rho_{k+1} \mid \mathcal{F}_{t_k}\right) + \mathbf{E}\left(z_{k+1}^3\rho_{k+1} \mid \mathcal{F}_{t_k}\right).\end{aligned} \tag{5.21}$$

By (1.5) we have

$$\begin{aligned}\left|\mathbf{E}\left(\eta_k^3\rho_{k+1} \mid \mathcal{F}_{t_k}\right)\right| &\le \left|\eta_k^3\mathbf{E}\left(\rho_{k+1} \mid \mathcal{F}_{t_k}\right)\right|\\ &\le \left|\eta_k\right|^3\left|\mathbf{E}\left(\rho_{k+1} \mid \mathcal{F}_{t_k}\right)\right|\\ &\le K\left(1+\overline{X}_k^2\right)^{1/2}\left|\eta_k\right|^3 h^{p_1}\\ &\le K\left(1+\left|\overline{X}_k\right|^4\right)^{1/4}\left|\eta_k\right|^3 h^{p_1}.\end{aligned} \tag{5.22}$$

Further (see (5.19) and (1.6))

$$\begin{aligned}\left|\mathbf{E}\left(\eta_k^2 z_{k+1}\rho_{k+1} \mid \mathcal{F}_{t_k}\right)\right| &\le \eta_k^2\left(\mathbf{E}\left(z_{k+1}^2 \mid \mathcal{F}_{t_k}\right)\right)^{1/2}\left(\mathbf{E}\left(\rho_{k+1}^2 \mid \mathcal{F}_{t_k}\right)\right)^{1/2}\\ &\le \eta_k^2 K\left|\eta_k\right| h^{1/2}\left(1+\left|\overline{X}_k\right|^2\right)^{1/2} h^{p_2}\\ &\le K\left(1+\left|\overline{X}_k\right|^4\right)^{1/4}\left|\eta_k\right|^3 h^{p_2+1/2}.\end{aligned} \tag{5.23}$$

Since

$$|\mathbf{E}(\eta_k z_{k+1} \cdot \eta_k \rho_{k+1} \mid \mathcal{F}_{t_k})| \leq \left(\mathbf{E}\left(\eta_k^2 z_{k+1}^2 \mid \mathcal{F}_{t_k}\right)\right)^{1/2} \left(\mathbf{E}\left(\eta_k^2 \rho_{k+1}^2 \mid \mathcal{F}_{t_k}\right)\right)^{1/2}$$
$$= |\eta_k| \left(\mathbf{E}\left(z_{k+1}^2 \mid \mathcal{F}_{t_k}\right)\right)^{1/2} |\eta_k| \left(\mathbf{E}\left(\rho_{k+1}^2 \mid \mathcal{F}_{t_k}\right)\right)^{1/2},$$

the third term at the righthand side of (5.21) can be estimated as in (5.23). The fourth and fifth terms can also be similarly estimated. For example,

$$|\mathbf{E}(\eta_k z_{k+1} \cdot z_{k+1} \rho_{k+1} \mid \mathcal{F}_{t_k})| \leq \mathbf{E}\left(|\eta_k| \cdot z_{k+1}^2 \cdot |\rho_{k+1}| \mid \mathcal{F}_{t_k}\right)$$
$$\leq |\eta_k| \left(\mathbf{E}\left(z_{k+1}^4 \mid \mathcal{F}_{t_k}\right)\right)^{1/2} \left(\mathbf{E}\left(\rho_{k+1}^2 \mid \mathcal{F}_{t_k}\right)\right)^{1/2}$$
$$\leq K\left(1 + \left|\overline{X}_k\right|^4\right)^{1/4} |\eta_k|^3 h^{p_2+1}. \tag{5.24}$$

Finally (see (5.2)):

$$\left|\mathbf{E}\left(z_{k+1}^3 \rho_{k+1} \mid \mathcal{F}_{t_k}\right)\right| \leq \left|\mathbf{E}\left(z_{k+1}^2 \cdot |z_{k+1}| \cdot |\rho_{k+1}| \mid \mathcal{F}_{t_k}\right)\right|$$
$$\leq \left(\mathbf{E}\left(z_{k+1}^4 \mid \mathcal{F}_{t_k}\right)\right)^{1/2} \left(\mathbf{E}\left(z_{k+1}^2 \rho_{k+1}^2 \mid \mathcal{F}_{t_k}\right)\right)^{1/2}$$
$$\leq K \eta_k^2 h \mathbf{E}\left(z_{k+1}^4 \mid \mathcal{F}_{t_k}\right)^{1/4} \mathbf{E}\left(\rho_{k+1}^4 \mid \mathcal{F}_{t_k}\right)^{1/4}$$
$$\leq K\left(1 + \left|\overline{X}_k\right|^4\right)^{1/4} |\eta_k|^3 h^{p_2+5/4}. \tag{5.25}$$

As a result we find

$$\left|\mathbf{E}\left(\alpha_{k+1}^3 \rho_{k+1} \mid \mathcal{F}_{t_k}\right)\right| \leq K\left(1 + \left|\overline{X}_k\right|^4\right)^{1/4} |\eta_k|^3 h^{p_2+1/2}.$$

Using the Hölder inequality and Lemma 5.2, we obtain

$$\left|\mathbf{E}\left(\alpha_{k+1}^3 \rho_{k+1}\right)\right| = \left|\mathbf{E}\mathbf{E}\left(\alpha_{k+1}^3 \rho_{k+1} \mid \mathcal{F}_{t_k}\right)\right| \leq \mathbf{E}\left|\mathbf{E}\left(\alpha_{k+1}^3 \rho_{k+1} \mid \mathcal{F}_{t_k}\right)\right|$$
$$\leq K\left(1 + \mathbf{E}\left|\overline{X}_k\right|^4\right)^{1/4} \cdot \left(\mathbf{E}\eta_k^4\right)^{3/4} h^{p_2+1/2}$$
$$\leq K\left(1 + \mathbf{E}\left|\overline{X}_0\right|^4\right)^{1/4} \delta_k^3 h^{p_2+1/2}.$$

This proves Lemma 5.3. □

Since $(\alpha_{k+1}\rho_{k+1})^2 \leq \alpha_{k+1}^2 \rho_{k+1}^2$, we only have to estimate $\mathbf{E}\alpha_{k+1}^2 \rho_{k+1}^2$.

LEMMA 5.4.

$$\mathbf{E}\alpha_{k+1}^2 \rho_{k+1}^2 \leq K\left(1 + \mathbf{E}\left|\overline{X}_0\right|^4\right)^{1/2} \delta_k^2 h^{2p_2}. \tag{5.26}$$

PROOF. We have

$$\mathbf{E}\left(\alpha_{k+1}^2 \rho_{k+1}^2 \mid \mathcal{F}_{t_k}\right) = \mathbf{E}\left(\eta_k^2 \rho_{k+1}^2 \mid \mathcal{F}_{t_k}\right) + 2\mathbf{E}\left(\eta_k z_{k+1} \rho_{k+1}^2 \mid \mathcal{F}_{t_k}\right)$$
$$+ \mathbf{E}\left(z_{k+1}^2 \rho_{k+1}^2 \mid \mathcal{F}_{t_k}\right). \tag{5.27}$$

Further:

$$\left|\mathbf{E}\left(\eta_k^2\rho_{k+1}^2 \mid \mathcal{F}_{t_k}\right)\right| \le \eta_k^2\mathbf{E}\left(\rho_{k+1}^2 \mid \mathcal{F}_{t_k}\right) \le K\left(1+\left|\overline{X}_k\right|^4\right)^{1/2}\eta_k^2 h^{2p_2}. \tag{5.28}$$

Then, using (1.11) and (5.2):

$$\begin{aligned}\left|\mathbf{E}\left(\eta_k z_{k+1}\rho_{k+1}^2 \mid \mathcal{F}_{t_k}\right)\right| &\le |\eta_k|\left(\mathbf{E}\left(z_{k+1}^2 \mid \mathcal{F}_{t_k}\right)\right)^{1/2}\cdot\left(\mathbf{E}\left(\rho_{k+1}^4 \mid \mathcal{F}_{t_k}\right)\right)^{1/2} \\ &\le K\eta_k^2\left(1+\left|\overline{X}_k\right|^4\right)^{1/2} h^{2p_2}.\end{aligned} \tag{5.29}$$

Finally,

$$\begin{aligned}\mathbf{E}\left(z_{k+1}^2\rho_{k+1}^2 \mid \mathcal{F}_{t_k}\right) &\le \left(\mathbf{E}\left(z_{k+1}^4 \mid \mathcal{F}_{t_k}\right)\right)^{1/2}\left(\mathbf{E}\left(\rho_{k+1}^4 \mid \mathcal{F}_{t_k}\right)\right)^{1/2} \\ &\le K\eta_k^2 h\left(1+\left|\overline{X}_k\right|^4\right)^{1/2} h^{2p_2-1/2}.\end{aligned} \tag{5.30}$$

It is readily seen that (5.26) follows from (5.27)–(5.30). □

LEMMA 5.5.

$$\left|\mathbf{E}\left(\alpha_{k+1}\rho_{k+1}^3\right)\right| \le K\left(1+\mathbf{E}\left|\overline{X}_0\right|^4\right)^{1/2}\delta_k^2 h^{2p_2} + K\left(1+\mathbf{E}\left|\overline{X}_0\right|^4\right)h^{4p_2-1}. \tag{5.31}$$

PROOF. By Lemma 5.4 and condition (5.2),

$$\begin{aligned}\left|\mathbf{E}\left(\alpha_{k+1}\rho_{k+1}^3\right)\right| &\le \mathbf{E}\left|\alpha_{k+1}\rho_{k+1}\right|\rho_{k+1}^2 \le \left(\mathbf{E}\alpha_{k+1}^2\rho_{k+1}^2\right)^{1/2}\left(\mathbf{E}\rho_{k+1}^4\right)^{1/2} \\ &\le K\left(1+\mathbf{E}\left|\overline{X}_0\right|^4\right)^{1/4}\delta_k h^{p_2}\left(1+\mathbf{E}\left|\overline{X}_0\right|^4\right)^{1/2} h^{2p_2-1/2}.\end{aligned}$$

This implies (5.31). □

PROOF OF THEOREM 5.1. Using the assumption (5.2), inequality (5.18), and Lemma 5.3, Lemma 5.4, Lemma 5.5, we obtain from (5.16):

$$\begin{aligned}\delta_{k+1}^4 &\le \delta_k^4(1+Kh) + 4K\left(1+\mathbf{E}\left|\overline{X}_0\right|^4\right)^{1/4}\delta_k^3 h^{p_2+1/2} \\ &\quad + K\left(1+\mathbf{E}\left|\overline{X}_0\right|^4\right)^{1/2}\delta_k^2 h^{2p_2} + K\left(1+\mathbf{E}\left|\overline{X}_0\right|^4\right)h^{4p_2-1} \\ &\le \delta_k^4(1+Kh) + K\left(1+\mathbf{E}\left|\overline{X}_0\right|^4\right)h^{4p_2-1}.\end{aligned}$$

By Lemma 1.3, for all $k = 0, \ldots, N$, this implies, since $\delta_0 = 0$,

$$\delta_k^4 \le K\left(1+\mathbf{E}\left|\overline{X}_0\right|^4\right)h^{4p_2-2}.$$

This proves Theorem 5.1. □

5.2. Construction of an auxiliary submartingale.

LEMMA 5.6. *Under the conditions of Theorem* 5.1 *the following inequality holds:*

$$\mathbf{E}\left(\max_{0\le k\le N}\left|\overline{X}_k\right|^4 \mid \mathcal{F}_{t_0}\right) \le K\left(1+|X_0|^4\right). \tag{5.32}$$

PROOF. We start with the well-known inequality (see [21, p. 121]):

$$\mathbf{E}\left(\max_{t_0\le t\le t_0+T}|X_{t_0,X_0}(t)|^4 \mid \mathcal{F}_{t_0}\right) \le K\left(1+|X_0|^4\right). \tag{5.33}$$

We further have

$$\left|\overline{X}_k\right|^4 = |X(t_k)+\eta_k| \le 8\,|X(t_k)|^4 + 8|\eta_k|^4.$$

Therefore

$$\max_{1\le k\le N}\left|\overline{X}_k\right|^4 \le 8\max_{1\le k\le N}|X(t_k)|^4 + 8\sum_{k=1}^N|\eta_k|^4.$$

Theorem 5.1 and (5.33) now imply

$$\mathbf{E}\left(\max_{1\le k\le N}\left|\overline{X}_k\right|^4 \mid \mathcal{F}_{t_0}\right) \le 8K\left(1+|X_0|^4\right) + K\left(1+|X_0|^4\right)\sum_{k=1}^N h^{2p_2-2}.$$

Since $p_2 \ge 3/4$ this implies (5.32). □

The following Lemma can be proved completely similarly.

LEMMA 5.7. *Under the conditions of Theorem* 1.1 *the following inequality holds:*

$$\mathbf{E}\left(\max_{0\le k\le N}\left|\overline{X}_k\right|^2 \mid \mathcal{F}_{t_0}\right) \le K\left(1+|X_0|^2\right). \tag{5.34}$$

This inequality is, of course, a simple consequence of (5.32), but the conditions in Lemma 5.7 are weaker than those in Lemma 5.6.

LEMMA 5.8. *Under the conditions of Theorem* 1.1 *the following inequality holds:*

$$\mathbf{E}\left(\eta_{k+1}^2 \mid \mathcal{F}_{t_k}\right) \ge \eta_k^2(1-Kh) - K\left(1+\overline{X}_k^2\right)h^{2p_2}. \tag{5.35}$$

PROOF. Write out

$$\begin{aligned}
\mathbf{E}\left(\eta_{k+1}^2 \mid \mathcal{F}_{t_k}\right) &= \mathbf{E}\left(\left(X_{t_k,X(t_k)}(t_{k+1}) - X_{t_k,\overline{X}_k}(t_{k+1})\right)^2 \mid \mathcal{F}_{t_k}\right) \\
&\quad + \mathbf{E}\left(\left(X_{t_k,\overline{X}_k}(t_{k+1}) - \overline{X}_{t_k,\overline{X}_k}(t_{k+1})\right)^2 \mid \mathcal{F}_{t_k}\right) \\
&\quad + 2\mathbf{E}\left(\left(X_{t_k,X(t_k)}(t_{k+1}) - X_{t_k,\overline{X}_k}(t_{k+1})\right)\right. \\
&\qquad \times \left.\left(X_{t_k,\overline{X}_k}(t_{k+1}) - \overline{X}_{t_k,\overline{X}_k}(t_{k+1})\right) \mid \mathcal{F}_{t_k}\right).
\end{aligned} \tag{5.36}$$

We have

$$\mathbf{E}\left(\left(X_{t_k,X(t_k)}(t_{k+1}) - X_{t_k,\overline{X}_k}(t_{k+1})\right)^2 \mid \mathcal{F}_{t_k}\right) = \eta_k^2 + \rho, \tag{5.37}$$

where

$$|\rho| \le \eta_k^2 \cdot Kh. \tag{5.38}$$

The inequality (5.38) and equation (5.37) readily follow from the proof of Lemma 1.1 if we use the first computation in this proof and then the inequalities (1.12) and (1.13). By (5.37) and (5.38) we obtain

$$\mathbf{E}\left(\left(X_{t_k,X(t_k)}(t_{k+1}) - X_{t_k,\overline{X}_k}(t_{k+1})\right)^2 \mid \mathcal{F}_{t_k}\right) \ge \eta_k^2(1-Kh). \tag{5.39}$$

Further,

$$\mathbf{E}\left(\left(X_{t_k,\overline{X}_k}(t_{k+1}) - \overline{X}_{t_k,\overline{X}_k}(t_{k+1})\right)^2 \mid \mathcal{F}_{t_k}\right) \le K\left(1+\left|\overline{X}_k\right|^2\right)\cdot h^{2p_2}. \tag{5.40}$$

It remains to give an upper bound for the last term in (5.36). By Lemma 1.1,

$$X_{t_k,X(t_k)}(t_{k+1}) - X_{t_k,\overline{X}_k}(t_{k+1}) = \eta_k + Z,$$
$$\mathbf{E}\left(Z^2 \mid \mathcal{F}_{t_k}\right) \le K\eta_k^2 h.$$

Using (1.5) and Theorem 1.1 we now estimate

$$\begin{aligned}
&\left|\mathbf{E}\left(\left(X_{t_k,X(t_k)}(t_{k+1}) - X_{t_k,\overline{X}_k}(t_{k+1})\right)\left(X_{t_k,\overline{X}_k}(t_{k+1}) - \overline{X}_{t_k,\overline{X}_k}(t_{k+1})\right) \mid \mathcal{F}_{t_k}\right)\right| \\
&\quad \le \left|\eta_k \mathbf{E}\left(\left(X_{t_k,\overline{X}_k}(t_{k+1}) - \overline{X}_{t_k,\overline{X}_k}(t_{k+1})\right) \mid \mathcal{F}_{t_k}\right)\right| \\
&\qquad + \left|\mathbf{E}\left(Z\left(X_{t_k,\overline{X}_k}(t_{k+1}) - \overline{X}_{t_k,\overline{X}_k}(t_{k+1})\right) \mid \mathcal{F}_{t_k}\right)\right| \\
&\quad \le |\eta_k|\cdot K\left(1+\left|\overline{X}_k\right|^2\right)^{1/2} h^{p_1} + K\cdot|\eta_k|\cdot h^{1/2}\left(1+\left|\overline{X}_k\right|^2\right)^{1/2} h^{p_2} \\
&\quad \le K|\eta_k|\cdot\left(1+\left|\overline{X}_k\right|^2\right)^{1/2} h^{p_2+1/2} \\
&\quad \le \frac{1}{2}K\eta_k^2 h + \frac{1}{2}\left(1+\left|\overline{X}_k\right|^2\right) h^{2p_2}.
\end{aligned} \tag{5.41}$$

By (5.39)–(5.41), from (5.36) we obtain

$$\left|\mathbf{E}\left(\eta_{k+1}^2 \mid \mathcal{F}_{t_k}\right)\right| \ge \eta_k^2(1-Kh) - K\left(1+\left|\overline{X}_k\right|^2\right)h^{2p_2} - K\eta_k^2 h - \left(1+\left|\overline{X}_k\right|^2\right)h^{2p_2},$$

which coincides with (5.35). □

The inequality (5.35) testifies of the fact that the sequence η_k^2 is a submartingale, up to $O(h)$. A submartingale $\tilde{\eta}_k^2$ can be constructed in the form

$$\tilde{\eta}_k^2 = \eta_k^2(1+Ch) + D_k\left(1+\overline{Y}_{k-1}^2\right)h^{2p_2}, \tag{5.42}$$

where

$$\overline{Y}_k^2 = \max_{0\le i\le k} \overline{X}_i^2. \tag{5.43}$$

Lemma 5.7 now ensures that $\mathbf{E}\tilde{\eta}_k^2$ exists.

We choose C_k and D_k such that the sequence $\tilde{\eta}_k^2$ becomes a submartingale, i.e. such that

$$\mathbf{E}\left(\tilde{\eta}_{k+1}^2 \mid \mathcal{F}_{t_k}\right) \geq \tilde{\eta}_k^2. \tag{5.44}$$

To this end we use the inequality (5.35):

$$\begin{aligned}\mathbf{E}\left(\tilde{\eta}_{k+1}^2 \mid \mathcal{F}_{t_k}\right) &= (1 + C_{k+1}h)\mathbf{E}\left(\eta_{k+1}^2 \mid \mathcal{F}_{t_k}\right) + D_{k+1}\left(1 + \overline{Y}_k^2\right)h^{2p_2} \\ &\geq (1 + C_{k+1}h)(1 - Kh)\eta_k^2 - K(1 + C_{k+1}h)\left(1 + \overline{X}_k^2\right)h^{2p_2} \\ &\quad + D_{k+1}\left(1 + \overline{Y}_k^2\right)h^{2p_2}.\end{aligned} \tag{5.45}$$

We now need a quantity smaller than the righthand side of (5.45) and greater than $\tilde{\eta}_k^2$ (in that case (5.44) holds and $\tilde{\eta}_k^2$ will be a submartingale):

$$\begin{aligned}&(1 + C_{k+1}h)(1 - Kh)\eta_k^2 - K(1 + C_{k+1}h)\left(1 + \overline{Y}_k^2\right)h^{2p_2} + D_{k+1}\left(1 + \overline{Y}_k^2\right)h^{2p_2} \\ &\geq \eta_k^2(1 + C_k h) + D_k\left(1 + \overline{Y}_{k-1}^2\right)h^{2p_2}.\end{aligned} \tag{5.46}$$

The inequality (5.46) holds if

$$(1 + C_{k+1}h)(1 - Kh) \geq 1 + C_k h, \qquad D_{k+1} - K(1 + C_{k+1}h) \geq D_k. \tag{5.47}$$

Putting, e.g. (we assume h to be sufficiently small),

$$C_0 = D_0 = 0, \quad C_{k+1}\frac{K + C_k}{1 - Kh}, \qquad D_{k+1} = D_k + K(1 + C_{k+1}h), \tag{5.48}$$

we arrive at (5.47). Recall that $k = 0, \ldots, N$, $h = T/N$, while for each N the sequences C_k and D_k are different. Using (5.48) it is easy to prove that we can find constants A and B independent of N and such that $A_k = C_k h \leq A$, $B_k = D_k h \leq B$ for all sufficiently large N and for $k = 0, \ldots, N$. Thus, we have provide the following Lemma.

LEMMA 5.9. *For all sufficiently large N there exist sequences A_k and B_k, uniformly bounded with respect to N and $k = 0, \ldots, N$, such that $(\tilde{\eta}_k^2, \mathcal{F}_{t_k})$, $k = 1, \ldots, N$, is a submartingale, where*

$$\tilde{\eta}_k^2 = \eta_k^2(1 + A_k) + B_k\left(1 + \overline{Y}_{k-1}^2\right)h^{2p_2-1}. \tag{5.49}$$

5.3. The strenghtened convergence theorem.

THEOREM 5.2. *Suppose that the conditions of Theorem 5.1 hold, i.e. the inequalities (1.5), (1.6), (5.2) hold and $p_2 \geq 3/4$, $p_1 \geq p_2 + 1/2$. Then*

$$\left(\mathbf{E}\max_{0\leq k\leq N}\left|X_{t_0,X_0}(t_k) - \overline{X}_{t_0,X_0}(t_k)\right|^2\right)^{1/2} \leq K\left(1 + \mathbf{E}|X_0|^4\right)^{1/4} h^{p_2-1/2}. \tag{5.50}$$

PROOF. First of all we prove that $\mathbf{E}\tilde{\eta}_N^4 < \infty$. In fact, (5.46) implies (recall that $\overline{Y}_{N-1}^2 \le \overline{Y}_N^2 = \max_{0\le k\le N} \overline{X}_k^2$, $A_k \le A$, $B_k \le B$)

$$\tilde{\eta}_N^4 \le 2(1+A)^2\eta_N^4 + 4B^2h^{4p_2-2} + 4B^2\overline{Y}_N^4 h^{4p_2-2}. \tag{5.51}$$

Theorem 5.1 implies

$$\mathbf{E}\eta_N^4 \le K\left(1+\mathbf{E}|X_0|^4\right)h^{4p_2-2}. \tag{5.52}$$

Further (since $\overline{Y}_N^4 = \left(\max_{0\le k\le N}\overline{X}_k^2\right)^2 = \max_{0\le k\le N}\overline{X}_k^4$), by Lemma 5.6 we can write

$$\mathbf{E}Y_N^4 \le K\left(1+\mathbf{E}|X_0|^4\right). \tag{5.53}$$

Finally, (5.51)–(5.53) imply

$$\mathbf{E}\tilde{\eta}_N^4 \le K\left(1+\mathbf{E}|X_0|^4\right)h^{4p_2-2}. \tag{5.54}$$

Since $\tilde{\eta}_k^2$ is a submartingale and $\mathbf{E}\tilde{\eta}_N^4 < \infty$, we have (see [23, p. 49])

$$\mathbf{E}\max_{1\le k\le N}\tilde{\eta}_k^4 \le 4\mathbf{E}\tilde{\eta}_N^4. \tag{5.55}$$

The inequalities $\eta_k^2 \le \tilde{\eta}_k^2$, (5.55) and (5.54) imply

$$\mathbf{E}\left(\max_{1\le k\le N}\eta_k^2\right) \le \mathbf{E}\left(\max_{1\le k\le N}\tilde{\eta}_k^2\right) \le \left(\mathbf{E}\left(\max_{1\le k\le N}\tilde{\eta}_k^2\right)^2\right)^{1/2}$$
$$= \left(\mathbf{E}\max_{1\le k\le N}\tilde{\eta}_k^4\right)^{1/2} \le 2\left(\mathbf{E}\tilde{\eta}_N^4\right)^{1/2} \le K\left(1+\mathbf{E}|X_0|^4\right)^{1/2}h^{2p_2-1}.$$

This proves the Theorem. □

CHAPTER 2

Modeling of Itô integrals

In the numerical integration formulas used in the Taylor-type expansion of solutions of systems of stochastic equations (see §2) the repeated *Itô integrals*

$$I_{i_1,\ldots,i_j} = \int_t^{t+h} dw_{i_j}(\theta) \int_t^{\theta} dw_{i_{j-1}}(\theta_1) \int_t^{\theta_1} \cdots \int_t^{\theta_{j-2}} dw_{i_1}(\theta_{j-1})$$

appeared, where $i_1, \ldots, i_j$ take values from the set $\{0, 1, \ldots, q\}$, and $dw_0(\theta_r)$ is understood to mean $d\theta_r$.

The *order* (of smallness) of the integral $I_{i_1,\ldots,i_j}$ is $\sum_{k=1}^{j}(2-\bar{\imath}_k)/2$ (see Lemma 2.1). A method of integral order m (see Theorem 2.1) includes all integrals $I_{i_1,\ldots,i_j}$ of order m and below; a method of half-integral order $m+1/2$ includes all integrals of order $m+1/2$ and below and also a deterministic integral of order $m+1$, which is not essential in the modeling. Thus, when using approximate numerical integration methods in practice, at each step there arises the difficult problem of modeling a certain set of Itô integrals. Since some of these can be expressed in terms of others, to start with it would be convenient to distinguish a possibly smallest set of variables from which all the required integrals could be found. As a result, the problem of modeling Itô integrals reduces to the modeling (exactly or approximately) of the distinguished random variables, which in general, of course, is an independent and difficult problem in its own right.

In this Chapter we undertake several approaches to solving these problems, and we will obtain a number of constructive results.

6. Modeling Itô integrals depending on a single noise

Such integrals arise in numerical integration formulas when the system of equations has the form

$$dX = a(t, X)\,dt + \sigma(t, X)\,dw, \tag{6.1}$$

with $w(t)$ a scalar Wiener process ($q = 1$).

6.1. Auxiliary formulas for single Itô integrals. Below we will consider integrals of the form

$$I(s) = \int_0^s \theta^k w^l(\theta)\, dw_i(\theta), \qquad i = 0, 1, \tag{6.2}$$

where $dw_0(\theta) = d\theta$, $dw_1(\theta) = dw(\theta)$.

The mean-square order of smallness of the integral (6.2) with respect to s as $s \to 0$ is equal to $m = k + (l/2) + (2 - i)/2$. We will call such integrals *m-integrals*. For example, $\int_0^s dw(\theta) = w(s)$ is a 1/2-integral, $\int_0^s d\theta = s$ is a 1-integral, $\int_0^s w(\theta)\, dw(\theta)$ is a 1-integral, etc. It is clear that the product of r integrals of orders $m_1, \dots, m_r$, respectively, has order $m_1 + \cdots + m_r$. Always, the order is either an integer or a half-integer.

We introduce the *Hermite polynomials*

$$H_n(t, x) = \frac{(-t)^n}{n!} \exp\left(\frac{x^2}{2t}\right) \frac{\partial^n}{\partial x^n} \exp\left(-\frac{x^2}{2t}\right), \qquad n \geq 0. \tag{6.3}$$

It can be immediately verified that

$$H_0 = 1, \qquad H_1 = x, \qquad H_n = \frac{xH_{n-1}}{n} - \frac{tH_{n-2}}{n}, \quad n \geq 2, \tag{6.4}$$

$$\frac{\partial H_n}{\partial x} = H_{n-1}, \qquad \frac{\partial H_n}{\partial t} + \frac{1}{2}\frac{\partial^2 H_n}{\partial x^2} = 0. \tag{6.5}$$

Consider $H_n(t, w(t))$ and compute, using Itô's formula and (6.5), the following differential:

$$dH_n = \left(\frac{\partial H_n}{\partial t} + \frac{1}{2}\frac{\partial^2 H_n}{\partial x^2}\right) dt + \frac{\partial H_n}{\partial x}\, dw = H_{n-1}\, dw. \tag{6.6}$$

Since $H_0 = 1$, we hence obtain a formula which is an instance of the *Itô–Wiener formula* (see [24]):

$$\int_0^t dw(\theta_1) \int_0^{\theta_1} dw(\theta_2) \cdots \int_0^{\theta_{n-1}} dw(\theta_n) = H_n(t, w(t)). \tag{6.7}$$

To find relations between integrals of the form (6.2) and of the same order, we consider the function $H_n(\alpha s, \beta w(s))$ and its Itô differential. By (6.5) and the equation

$\partial^2 H_n/\partial x^2 = H_{n-2}$ we have

$$\begin{aligned} dH_n(\alpha s, \beta w(s)) &= \left(\alpha \frac{\partial H_n}{\partial t}(\alpha s, \beta w(s)) + \frac{1}{2}\beta^2 \frac{\partial^2 H_n}{\partial x^2}(\alpha s, \beta w(s))\right) ds \\ &\quad + \beta \frac{\partial H_n}{\partial x}(\alpha s, \beta w(s))\, dw(s) \\ &= \frac{1}{2}(\beta^2 - \alpha) \frac{\partial^2 H_n}{\partial x^2}(\alpha s, \beta w(s))\, ds + \beta H_{n-1}(\alpha s, \beta w(s))\, dw(s) \\ &= \frac{1}{2}(\beta^2 - \alpha) H_{n-2}(\alpha s, \beta w(s))\, ds + \beta H_{n-1}(\alpha s, \beta w(s))\, dw(s). \end{aligned} \tag{6.8}$$

Whence,

$$\begin{aligned} H_n(\alpha s, \beta w(s)) &= \frac{1}{2}(\beta^2 - \alpha) \int_0^s H_{n-2}(\alpha\theta, \beta w(\theta))\, d\theta \\ &\quad + \beta \int_0^s H_{n-1}(\alpha\theta, \beta w(\theta))\, dw(\theta). \end{aligned} \tag{6.9}$$

We write out the first eight Hermite polynomials:

$$\begin{gathered} H_0 = 1, \qquad H_1 = x, \qquad H_2 = \frac{x^2}{2} - \frac{t}{2}, \qquad H_3 = \frac{x^3}{6} - \frac{tx}{2}, \qquad H_4 = \frac{x^4}{24} - \frac{x^2 t}{4} + \frac{t^2}{8}, \\ H_5 = \frac{x^5}{120} - \frac{x^3 t}{12} + \frac{x t^2}{8}, \qquad H_6 = \frac{x^6}{720} - \frac{x^4 t}{48} + \frac{x^2 t^2}{16} - \frac{t^3}{48}, \\ H_7 = \frac{x^7}{7 \cdot 720} - \frac{x^5 t}{240} + \frac{x^3 t^2}{48} - \frac{x t^3}{48}. \end{gathered} \tag{6.10}$$

Writing down (6.9) for $n = 2, \dots, 7$ in succession and equating the expressions at identical powers of α and β, we obtain the required relations between the $n/2$-integrals.

For $n = 2$ relation (6.9) gives

$$-\frac{\alpha s}{2} + \frac{\beta^2 w^2(s)}{2} = \frac{1}{2}(\beta^2 - \alpha)s + \beta \int_0^s \beta w(\theta)\, dw(\theta).$$

Hence,

$$\int_0^s w(\theta)\, dw(\theta) = \frac{w^2(s) - s}{2}. \tag{6.11}$$

For $n = 3$:

$$\begin{aligned} -\frac{\alpha\beta w(s)}{2} + \frac{\beta^3 w^3(s)}{6} &= \frac{1}{2}\beta^3 \int_0^s w(\theta)\, d\theta - \frac{1}{2}\alpha\beta \int_0^s w(\theta)\, d\theta \\ &\quad + \frac{\beta^3}{2} \int_0^s w^2(\theta)\, d\theta - \frac{\alpha\beta}{2} \int_0^s \theta\, dw(\theta). \end{aligned}$$

This implies the following formulas for the 3/2-integrals:

$$\int_0^s \theta\, dw(\theta) = sw(s) - \int_0^s w(\theta)\, d\theta, \tag{6.12}$$

$$\int_0^s w^2\, dw = \frac{w^3(s)}{3} - \int_0^s w(\theta)\, d\theta. \tag{6.13}$$

Proceeding in this way, we obtain for $n = 4$ the following formulas for the 2-integrals:

$$\int_0^s \theta w(\theta)\, dw = -\frac{s^2}{4} + \frac{w^2(s)\cdot s}{2} - \frac{1}{2}\int_0^s w^2(\theta)\, d\theta, \tag{6.14}$$

$$\int_0^s w^3(\theta)\, dw = \frac{w^4(s)}{4} - \frac{3}{2}\int_0^s w^2(\theta)\, d\theta. \tag{6.15}$$

And for $n = 5$ the following formulas for the 5/2-integrals:

$$\int_0^s \theta^2\, dw = s^2 w(s) - 2\int_0^s \theta w(\theta)\, d\theta, \tag{6.16}$$

$$\int_0^s \theta w^2(\theta)\, dw = \frac{1}{3} s w^3(s) - \int_0^s \theta w(\theta)\, d\theta - \frac{1}{3}\int_0^s w^3(\theta)\, d\theta, \tag{6.17}$$

$$\int_0^s w^4(\theta)\, dw = \frac{1}{5} w^5(s) - 2\int_0^s w^3(\theta)\, d\theta. \tag{6.18}$$

Comparison of the cases for $n = 2, 3, 4, 5$ shows that when modeling all single 1-integrals (all in all one integral) we do not have to model anything else but $w(s)$; when modeling all single 3/2- and 2-integrals, in addition we have to model one integral in each case ($\int_0^s w(\theta)\, d\theta$ and $\int_0^s w^2(\theta)\, d\theta$, respectively); when modeling all single 5/2-integrals, in addition we have to model already two integrals ($\int_0^s \theta w(\theta)\, d\theta$ and $\int_0^s w^3(\theta)\, d\theta$, respectively).

Of course, there arises the natural question whether, e.g., the 5/2-integral $\int_0^s \theta w(\theta)\, d\theta$ can be expressed in terms of $\int_0^s w^3(\theta)\, d\theta$ and a combination of products of 1/2- and 2- or 1- and 3/2-integrals. We have sufficiently convincing reasons (which we do not give here) supporting the fact that this cannot be done; however, we do not have a strict proof of this fact.

For $n = 6$ we have

$$\int_0^s \theta^2 w(\theta)\,dw = \frac{1}{2}s^2 w^2(s) - \frac{1}{6}s^3 - \int_0^s \theta w^2(\theta)\,d\theta, \tag{6.19}$$

$$\int_0^s \theta w^3(\theta)\,dw = \frac{1}{4}s w^4(s) - \frac{1}{4}\int_0^s w^4(\theta)\,d\theta - \frac{3}{2}\int_0^s \theta w^2(\theta)\,d\theta, \tag{6.20}$$

$$\int_0^s w^5(\theta)\,dw = \frac{1}{6}w^6(s) - \frac{5}{2}\int_0^s w^4(\theta)\,d\theta. \tag{6.21}$$

Again, in addition we have to model two integrals. One can get convinced of the fact that when modeling all 7/2-integrals, only three integrals have to be modeled in addition. We restrict ourselves to the formulas written out above. We draw attention to the fact that all m-integrals, $m = 1/2, 1, 3/2, 2, 5/2, 3$, with respect to $dw(\theta)$ can be expressed in terms of m-integrals with respect to $d\theta$. We also note that all formulas (6.11)–(6.21) can be readily verified by differentiating both sides of the equations.

6.2. Reduction of repeated Itô integrals to single Itô integrals. We will successively transform repeated integrals of order $1/2, 1, 3/2$, etc. Here, the indices $i_1, \dots, i_j$ in the integral $I_{i_1 \dots, i_j}$ can take either the value 0 (then $dw_0(\theta) = d\theta$) or 1 (then $dw_1(\theta) = dw(\theta)$). Without loss of generality we may take $t = 0$. In the sequel we will keep the notation

$$I_{i_1,\dots,i_j}(s) = \int_0^s dw_{i_j}(\theta) \int_0^\theta dw_{i_{j-1}}(\theta_1) \int_0^{\theta_1} \cdots \int_0^{\theta_{j-2}} dw_{i_1}(\theta_{j-1}) \tag{6.22}$$

also for repeated integrals of order m, with m an integer or a half-integer; we will call such integrals *m-integrals*, as in the case of a single integral.

We have

$$I_{i_1,\dots,i_j}(s) = \int_0^s I_{i_1,\dots,i_{j-1}}(\theta)\,dw_{i_j}(\theta). \tag{6.23}$$

This formula makes clear that to obtain all m-integrals of the form (6.22) we, first, have to consider all $(m-1)$-integrals $I_{i_1,\dots,i_{j-1}}$ and construct from them the m-integrals

$$I_{i_1,\dots,i_{j-1},0}(s) = \int_0^s I_{i_1,\dots,i_{j-1}}(\theta)\,d\theta, \tag{6.24}$$

and, secondly, have to consider all $(m - 1/2)$-integrals $I_{i_1,\dots,i_{j-1}}$ and construct from them the m-integrals

$$I_{i_1,\dots,i_{j-1},1}(s) = \int_0^s I_{i_1,\dots,i_{j-1}}(\theta)\,dw(\theta). \tag{6.25}$$

Guided by this rule we will successively construct the m-integrals, starting from the 1/2-integrals (there is only one such) and the 1-integrals (there are two of these). In fact, the 1/2-integral is

$$I_1(s) = \int_0^s dw(\theta) = w(s), \tag{6.26}$$

and the 1-integrals are

$$I_0(s) = \int_0^s d\theta = s, \qquad I_{1,1}(s) = \int_0^s I_1(\theta)\, dw(\theta) = \frac{w^2(s) - s}{2}. \tag{6.27}$$

Using the recurrence rule (6.23)–(6.25) and the formulas (6.12), (6.13), we write out the 3/2-integrals:

$$I_{1,0}(s) = \int_0^s w(\theta)\, d\theta, \qquad I_{0,1}(s) = \int_0^s \theta\, dw(\theta) = sw(s) - \int_0^s w(\theta)\, d\theta,$$

$$I_{1,1,1}(s) = \int_0^s \left(\frac{w^2(\theta) - v}{2}\right) dw(\theta) = \frac{w^3(s)}{6} - \frac{1}{2} sw(s). \tag{6.28}$$

We write out the 2-integrals. Using the 1-integrals (6.27) we have

$$I_{0,0} = \frac{s^2}{2}, \qquad I_{1,1,0}(s) = \frac{1}{2}\int_0^s w^2(\theta)\, d\theta - \frac{s^2}{4}, \tag{6.29}$$

and subsequently, using the 3/2-integrals (6.28) and relations (6.14), (6.15):

$$\begin{aligned} I_{1,0,1}(s) &= \int_0^s \left(\int_0^\theta w(\theta_1)\, d\theta_1\right) dw(\theta) \\ &= w(s) \int_0^s w(\theta)\, d\theta - \int_0^s w^2(\theta)\, d\theta, \end{aligned} \tag{6.30}$$

$$\begin{aligned} I_{0,1,1}(s) &= \int_0^s \theta w(\theta)\, dw(\theta) - \int_0^s \left(\int_0^\theta w(\theta_1)\, d\theta_1\right) dw(\theta) \\ &= -\frac{s^2}{4} + \frac{sw^2(s)}{2} - w(s)\int_0^s w(\theta)\, d\theta + \frac{1}{2}\int_0^s w^2(\theta)\, d\theta, \end{aligned} \tag{6.31}$$

$$I_{1,1,1,1}(s) = \int_0^s \left(\frac{w^3(\theta)}{6} - \frac{1}{2}\theta w(\theta)\right) dw(\theta) = \frac{s^2}{8} - \frac{sw^2(s)}{4} + \frac{1}{24} w^4(s). \tag{6.32}$$

So, as a result all five repeated 2-integrals can be expressed in terms of $w(s)$, $\int_0^s w(\theta)\, d\theta$, and $\int_0^s w^2(\theta)\, d\theta$.

We write out the 5/2-integrals. Using the 3/2-integrals (6.28) we have

$$I_{1,0,0} = \int_0^s \left(\int_0^\theta w(\theta_1)\, d\theta_1 \right) d\theta = s \int_0^s w(\theta)\, d\theta - \int_0^s \theta w(\theta)\, d\theta, \tag{6.33}$$

$$I_{0,1,0}(s) = 2 \int_0^s \theta w(\theta)\, d\theta - s \int_0^s w(\theta)\, d\theta, \tag{6.34}$$

$$I_{1,1,1,0}(s) = \frac{1}{6} \int_0^s w^3(\theta)\, d\theta - \frac{1}{2} \int_0^s \theta w(\theta)\, d\theta. \tag{6.35}$$

Using the 2-integrals (6.29)–(6.32) and relations (6.16)–(6.18):

$$I_{0,0,1}(s) = \frac{1}{2} \int_0^s \theta^2\, dw = \frac{1}{2} s^2 w(s) - \int_0^s \theta w(\theta)\, d\theta, \tag{6.36}$$

$$\begin{aligned} I_{1,1,0,1}(s) &= \frac{1}{2} \int_0^s \left(\int_0^\theta w^2(\theta_1)\, d\theta_1 \right) dw(\theta) - \int_0^s \frac{\theta^2}{4}\, dw(\theta) \\ &= \frac{1}{2} w(s) \int_0^s w^2(\theta)\, d\theta - \frac{1}{2} \int_0^s w^3(\theta)\, d\theta - \frac{1}{4} s^2 w(s) + \frac{1}{2} \int_0^s \theta w(\theta)\, d\theta, \end{aligned} \tag{6.37}$$

$$I_{1,0,1,1}(s) = \int_0^s \left(w(\theta) \int_0^\theta w(\theta_1)\, d\theta_1 \right) dw(\theta) - \int_0^s \left(\int_0^\theta w^2(\theta_1)\, d\theta_1 \right) dw(\theta).$$

Here we encounter integrals which can be transformed into a sum of products of single integrals in a way that, although not complicated, is nevertheless not trivial. In fact, we can immediately verify that

$$\begin{aligned} \int_0^s \left(w(\theta) \int_0^\theta w(\theta_1)\, d\theta_1 \right) dw(\theta) &= \frac{1}{2} w^2(s) \int_0^s w(\theta)\, d\theta - \frac{1}{2} \int_0^s w^3(\theta)\, d\theta \\ &\quad - \frac{1}{2} s \int_0^s w(\theta)\, d\theta + \frac{1}{2} \int_0^s \theta w(\theta)\, d\theta. \end{aligned}$$

As a result,

$$\begin{aligned} I_{1,0,1,1}(s) &= \frac{1}{2} w^2(s) \int_0^s w(\theta)\, d\theta + \frac{1}{2} \int_0^s w^3(\theta)\, d\theta - \frac{1}{2} s \int_0^s w(\theta)\, d\theta \\ &\quad + \frac{1}{2} \int_0^s \theta w(\theta)\, d\theta - w(s) \int_0^s w^2(\theta)\, d\theta. \end{aligned} \tag{6.38}$$

We do not know whether an arbitrary repeated m-integral $I_{i_1,\dots,i_j}(s)$ can be reduced to a sum of products of single integrals of the form (6.2). However, all repeated 5/2- and

3-integrals $I_{i_1,\dots,i_j}(s)$ can be reduced to such sums. We give a complete computation of the 5/2-integrals. We have:

$$I_{0,1,1,1}(s) = -\int_0^s \frac{\theta^2}{4}\,dw(\theta) + \frac{1}{2}\int_0^s \theta w^2(\theta)\,dw(\theta)$$
$$-\int_0^s \left(w(\theta)\int_0^\theta w(\theta_1)\,d\theta_1\right) dw(\theta) + \frac{1}{2}\int_0^s \left(\int_0^\theta w^2(\theta_1)\,d\theta_1\right) dw(\theta)$$
$$= -\frac{1}{4}s^2 w(s) + \frac{1}{6}sw^3(s) - \frac{1}{6}\int_0^s w^3(\theta)\,d\theta - \frac{1}{2}w^2(s)\int_0^s w(\theta)\,d\theta$$
$$+\frac{1}{2}s\int_0^s w(\theta)\,d\theta - \frac{1}{2}\int_0^s \theta w(\theta)\,d\theta + \frac{1}{2}w(s)\int_0^s w^2(\theta)\,d\theta. \tag{6.39}$$

Finally (here it is easier to use (6.7), instead of (6.16)–(6.16)),

$$I_{1,1,1,1,1}(s) = \frac{1}{8}s^2 w(s) - \frac{1}{12}sw^3(s) + \frac{1}{120}w^5(s). \tag{6.40}$$

Now we can subsume the modeling of random variables for the system (6.1) with a single noise. To construct methods of order 1/2 or 1, it suffices at each step to model $w(h)$. To construct a method of order 3/2, it suffices to model at each step $w(h)$ and $\int_0^h w(\theta)\,d\theta$. Here, all repeated 3/2-integrals participating in the method can be found from (6.28). To construct a method of order 2, it suffices to model at each step the three random variables $w(h)$, $\int_0^h w(\theta)\,d\theta$, and $\int_0^h w^2(\theta)\,d\theta$. All repeated 2-integrals involved can be found from (6.29)–(6.32). To construct a method of order 5/2, it suffices to model at each step the five random variable $w(h)$, $\int_0^h w(\theta)\,d\theta$, $\int_0^h w^2(\theta)\,d\theta$, $\int_0^h \theta w(\theta)\,d\theta$, and $\int_0^h w^3(\theta)\,d\theta$. All repeated 5/2-integrals involved can be found from (6.33)–(6.40). It is clear that methods of higher order require a yet larger expense as regards modeling the repeated Itô integrals involved.

6.3. Exact modeling of the random variables $w(h)$, $\int_0^h w(\theta)\,d\theta$, and $\int_0^h w^2(\theta)\,d\theta$. The problem of modeling $w(h)$ and $\int_0^h w(\theta)\,d\theta$ can be solved very simply. In fact, their joint distribution is Gaussian. Write the integral $\int_0^h w(\theta)\,d\theta$ as

$$\int_0^h w(\theta)\,d\theta = \alpha w(h) + \left(\int_0^h w(\theta)\,d\theta - \alpha w(h)\right),$$

and choose α such that $w(h)$ and $\int_0^h w(\theta)\,d\theta - \alpha w(h)$ are independent. Clearly, α can be found from the condition that the mathematical expectation of their product should vanish. Since $\mathbf{E}w(h)\int_0^h w(\theta)\,d\theta = h^2/2$, we have $\alpha = h/2$. As a result, the integral $\int_0^h w(\theta)\,d\theta$ can be written as a sum of two independent normally distributed random variables:

$$\int_0^h w(\theta)\,d\theta = \frac{1}{2}hw(h) + \left(\int_0^h w(\theta)\,d\theta - \frac{1}{2}hw(h)\right). \tag{6.41}$$

The first term at the righthand side of (6.41) is $N(0, h^3/4)$-distributed, and the second term is $N(0, h^3/12)$-distributed. Thus, from the point of view of modeling the random variables involved, a method of order 3/2 for a system (6.1) with a single noise is rather simple.

We turn to a method of order of accuracy 2. Here, in general, at each step we have to model the three random variables $w(h)$, $\int_0^h w(\theta)\,d\theta$, and $\int_0^h w^2(\theta)\,d\theta$ (see the previous Subsection). The density $p(s,x,y,z)$ of the distribution of these variables satisfies the *Kolmogorov equation*

$$\frac{\partial p}{\partial s} - \frac{1}{2}\frac{\partial^2 p}{\partial x^2} - x\frac{\partial p}{\partial y} - x^2\frac{\partial p}{\partial z} = 0, \qquad p(0,x,y,z) = \delta(x,y,z).$$

An analytic solution of this equation is hardly seen immediately. The search for the characteristic function $g(\lambda_1, \lambda_2, \lambda_3)$ of these variables reduces to the computation of the Wiener integral

$$I = \mathbf{E}e^{i\left(\lambda_1 w(h) + \int_0^h (\lambda_2 w(\theta) + \lambda_3 w^2(\theta))\,d\theta\right)} = g(\lambda_1, \lambda_2, \lambda_3),$$

which can be computed by using a simple generalisation of the *Cameron-Martin formula* (see [23, p. 323]). We give the necessary derivations. Consider the random variable

$$\chi(t) = \exp\left[i\left(p(t)w^2(t) + q(t)w(t) + r(t) + \int_0^t \left(\lambda_2 w(s) + \lambda_3 w^2(s)\right) ds\right)\right],$$

where the differentiable functions $p(t), q(t), r(t)$ are defined below. We have

$$\begin{aligned} d\chi = {} & i\chi\left(\left(p'w^2 + q'w + r' + \lambda_2 w + \lambda_3 w^2 + p\right) dt + (2pw + q)\,dw\right) \\ & - \frac{1}{2}\chi(2pw+q)^2\,dt. \end{aligned} \tag{6.42}$$

Suppose that

$$p(h) = 0, \qquad q(h) = \lambda_1, \qquad r(h) = 0. \tag{6.43}$$

Then

$$\mathbf{E}\chi(h) = g(\lambda_1, \lambda_2, \lambda_3). \tag{6.44}$$

Further, suppose that p, q, r are such that

$$i\left(p'w^2 + q'w + r' + \lambda_2 w + \lambda_3 w^2 + p\right) - 2p^2w^2 - 2pqw - \frac{q^2}{2} = 0$$

on the interval $[0, h]$. For this it suffices that

$$p' + \lambda_3 + 2ip^2 = 0, \qquad q' + \lambda_2 + 2ipq = 0, \qquad r' + p + i\frac{q^2}{2} = 0. \tag{6.45}$$

Suppose, conversely, that for all real $\lambda_1, \lambda_2, \lambda_3$ there are functions $p(t), q(t), r(t)$, defined on $0 \le t \le h$, which are a solution of the Cauchy problem (6.45), (6.43). Then (6.42) implies

$$d\chi = i(2pw + q)\chi\, dw. \tag{6.46}$$

Since $\mathbf{E}|2pw + q|^2|\chi|^2$ exists and is finite, (6.46) implies

$$d\mathbf{E}\chi = 0$$

and, hence, $\mathbf{E}\chi = \text{const} = \mathbf{E}\chi(0) = e^{ir0)} = \mathbf{E}\chi(h) = g(\lambda_1, \lambda_2, \lambda_3)$.

Thus, we have obtained the following result (it does not immediately follow from this version of the Cameron–Martin formula as, e.g., given in [23], but nevertheless this result can be regarded as an instance of a more general formula related to the Cameron–Martin formula).

LEMMA 6.1. *The characteristic function $g(\lambda_1, \lambda_2, \lambda_3)$ of the random variables $w(h)$, $\int_0^h w(s)\, ds$, $\int_0^h w^2(s)\, ds$ is equal to*

$$g(\lambda_1, \lambda_2, \lambda_3) = e^{ir(0)},$$

where $r(0)$ is the value at $t = 0$ of the coordinate $r(t)$ of the solution $(p(t), q(t), r(t))$ of the system (6.45) satisfying the initial data (6.43).

Of course, $r(0)$ can be computed. As a result we obtain

$$\begin{aligned} g(\lambda_1, \lambda_2, \lambda_3) &= e^{ir(0)} \\ &= \frac{1}{\sqrt{\cosh 2ip_0h}} \exp\left\{ -\frac{i\lambda_2^2}{4\lambda_3}\left(h + \frac{1}{2ip_0} th(-2ip_0h)\right) \right. \\ &\quad \left. - \frac{i\lambda_1^2}{4p_0} th(-2ip_0h) + \frac{i\lambda_1\lambda_2}{2\lambda_3}\left(\frac{1}{\cosh(2ip_0h)} - 1\right) \right\}, \end{aligned}$$

where $p_0 = \sqrt{\bar{\imath}\lambda_3/2}$.

It is seen that the characteristics of the distribution of the random variables under consideration are sufficiently complicated, and will hardly be useful in practical modeling problems. Thus, exact modeling has bad perspectives, and therefore there arises the need of approximately modeling these variables.

6.4. Approximate modeling of the random variables $w(h)$, $\int_0^h w(\theta)\, d\theta$, and $\int_0^h w^2(\theta)\, d\theta$.

LEMMA 6.2. *Suppose that the one-step approximation (see (1.3))*

$$\overline{X}_{t,x}(t+h) = x + A(t, x, h; w_i(\theta) - w_i(t),\ i = 1, \dots, q, t \le \theta \le t+h) \tag{6.47}$$

generates a method with order of accuracy m. Suppose A contains terms of the form $P(t, x) \cdot \xi(w_i(\theta) - w_i(t),\ i = 1, \dots, q,\ t \le \theta \le t+h)$, where $|P(t, x)| \le K\, (1 + |x|^2)^{1/2}$ and ξ is a random variable, depending on the Wiener processes on the interval $[t, t+h]$, *as indicated between the brackets. Let $\xi = \eta + \zeta$, where η and ζ are random variables depending on these very same Wiener processes on the same interval. Finally, suppose*

$$|\mathbf{E}\zeta| \le Kh^{m+1}, \qquad \left(\mathbf{E}\zeta^2\right)^{1/2} \le Kh^{m+1/2}. \tag{6.48}$$

Then the method based on the one-step approximation and with $P \cdot \xi$ replaced by $P \cdot \eta$ has order of accuracy equal to m.

Clearly, the proof of this Lemma follows from the main Theorem 1.1.

This Lemma makes it possible to replace random variables participating in the method and difficult to model by random variables for which this is simpler. A sufficiently general approach to the approximate modeling of random variables is as follows. Make up the system of stochastic differential equations whose solution at time h is given by the set of random variables to be modeled. If this system is integrable over the interval $[0, h]$ with sufficiently high order of accuracy (at the expense of a small integration step h_1, this can be done by a 'rough' method in which simpler random variables are modeled), then we can construct the required approximations for the variables to be modeled. We note one important particular point. It is a fact that the random variables to be modeled each have different orders of smallness with respect to h. And so we need approximations for them of a different order of accuracy. We precisise this using the variables $w(h)$, $\int_0^h w(\theta)\,d\theta$, and $\int_0^h w^2(\theta)\,d\theta$ as example.

Introduce the new process

$$v(s) = \frac{w(sh)}{\sqrt{h}}, \qquad 0 \le s \le 1. \tag{6.49}$$

It is obvious that $v(s)$ is a standard Wiener process. We have

$$w(h) = h^{1/2}v(1), \qquad \int_0^h w(\theta)\,d\theta = h^{3/2}\int_0^1 v(s)\,ds, \qquad \int_0^h w^2(\theta)\,d\theta = h^2\int_0^1 v^2(s)\,ds. \tag{6.50}$$

Thus, the problem of modeling the variables $w(h)$, $\int_0^h w(\theta)\,d\theta$, and $\int_0^h w^2(\theta)\,d\theta$ can be reduced to that of modeling the variables $v(1)$, $\int_0^1 v(s)\,ds$, and $\int_0^1 v^2(s)\,ds$. These variables are the solution of the system of equations

$$\begin{aligned} dx &= dv(s), & x(0) &= 0,\\ dy &= x\,ds, & y(0) &= 0,\\ dz &= x^2\,ds, & z(0) &= 0 \end{aligned} \tag{6.51}$$

at the moment $s = 1$.

Let $\overline{x}(s_k)$, $\overline{y}(s_k)$, $\overline{z}(s_k)$, $0 \le s_0 < s_1 < \cdots < s_N = 1$, $s_{k+1} - s_k = h_1 = 1/N$, be an approximate solution of (6.51). If we are guided by Lemma 6.1 when constructing a method of order of accuracy equal to 2, then the random variables $w(h) = h^{1/2}x(1)$, $\int_0^h w(\theta)\,d\theta = h^{3/2}y(1)$, and $\int_0^h w^2(\theta)\,d\theta = h^2 z(1)$ participating in it can be replaced by

$h^{1/2}\overline{x}(1)$, $h^{3/2}\overline{y}(1)$, and $h^2\overline{z}(1)$, if only the conditions

$$|\mathbf{E}\,(v(1) - \overline{x}(1))| = O(h^{5/2}), \qquad \left(\mathbf{E}\,(v(1) - \overline{x}(1))^2\right)^{1/2} = O(h^2), \tag{6.52}$$

$$\left|\mathbf{E}\left(\int_0^1 v(s)\,ds - \overline{y}(1)\right)^2\right| = O(h^{3/2}), \qquad \left(\mathbf{E}\left(\int_0^1 v(s)\,ds - \overline{y}(1)\right)^2\right)^{1/2} = O(h), \tag{6.53}$$

$$\left|\mathbf{E}\left(\int_0^1 v^2(s)\,ds - \overline{z}(1)\right)\right| = O(h), \qquad \left(\mathbf{E}\left(\int_0^1 v^2(s)\,ds - \overline{z}(1)\right)^2\right)^{1/2} = O(h^{1/2}). \tag{6.54}$$

We consider first the Euler method for integrating (6.51) with step h_1:

$$\begin{aligned} x_{k+1} &= x_k + \Delta_k v(h_1), & x_0 &= 0, \\ y_{k+1} &= y_k + x_k h_1, & y_0 &= 0, \\ z_{k+1} &= z_k + x_k^2 h_1, & z_0 &= 0. \end{aligned} \tag{6.55}$$

Since

$$\begin{aligned} x_k &= v(s_k), & x_N &= \overline{x}(1) = v(1), \\ y_k &= h_1 \sum_{i=0}^{k-1} v(s_i), & \overline{y}(1) &= h_1 \sum_{i=0}^{N-1} v(s_i), \\ z_k &= h_1 \sum_{i=0}^{k-1} v^2(s_i), & \overline{z}(1) &= h_1 \sum_{i=0}^{N-1} v^2(s_i), \end{aligned}$$

we have

$$\mathbf{E}v(1) = \mathbf{E}\overline{x}(1) = 0,$$

$$\mathbf{E}\int_0^1 v(s)\,ds = \mathbf{E}\overline{y}(1) = 0,$$

$$\mathbf{E}\int_0^1 v^2(s)\,ds = \frac{1}{2}, \qquad \mathbf{E}\overline{z}(1) = \frac{1}{2} - \frac{h_1}{2}. \tag{6.56}$$

Further, since the system (6.51) is a system with additive noises, Euler's method has order $O(h_1)$. Therefore, taking $h_1 = h$ we find that all relations (6.52)–(6.54) are fulfilled (note that in our case $\mathbf{E}\,(v(1) - \overline{x}(1))^2 = 0$). Immediate computations (of quite some length in the case of the third equation) give (for $h_1 = h$):

$$\left| \mathbf{E}\left(\int_0^1 v^2(s)\,ds - \overline{z}(1) \right) \right| = \frac{h}{2},$$

$$\mathbf{E}\left(\int_0^1 v(s)\,ds - \overline{y}(1) \right)^2 = \frac{h^2}{3},$$

$$\mathbf{E}\left(\int_0^1 v^2(s)\,ds - \overline{z}(1) \right)^2 = \frac{11}{12}h^2 - \frac{h^3}{3}. \tag{6.57}$$

Thus, if we approximately model the random variables by a method of order 2 (for a system with a single noise) by applying Euler's method to (6.51), then at each step we have to model $\approx 1/h$ normally distributed random variables (here, h is the integration step in both the initial as well as the auxiliary system (6.51)).

We will now use a method of order 3/2 (see (3.2)) for integrating (6.46). We have

$$a = \begin{bmatrix} 0 \\ x \\ x^2 \end{bmatrix}, \qquad \sigma = \begin{bmatrix} 1 \\ 0 \\ 0 \end{bmatrix}, \qquad \Lambda a = \begin{bmatrix} 0 \\ 1 \\ 2x \end{bmatrix}, \qquad La = \begin{bmatrix} 0 \\ 0 \\ 1 \end{bmatrix},$$

$$x_{k+1} = x_k + \Delta_k v(h_1),$$

$$y_{k+1} = y_k + x_k h_1 + \int_{s_k}^{s_{k+1}} (v(\theta) - v(s_k))\,d\theta,$$

$$z_{k+1} = z_k + x_k^2 h_1 + 2x_k \int_{s_k}^{s_{k+1}} (v(\theta) - v(s_k))\,d\theta + \frac{h_1^2}{2}. \tag{6.58}$$

The method (6.58) has the following properties. First, x_k and y_k are equal to $v(s_k)$ and $\int_0^{s_k} v(\theta)\,d\theta$, respectively (this is obvious), and, secondly, $\mathbf{E}\overline{z}(1) = \mathbf{E}z(1)$ (this will be proved below). As a result, out of all relations (6.52)–(6.54) only the second relation in (6.54) has to be satisfied. However, since the method (6.58) has order 3/2, this relation reduces to the requirement

$$\left(\mathbf{E}\,(z(1) - \overline{z}(1))^2\right)^{1/2} = O(h_1^{3/2}) = O(h^{1/2}).$$

Thus, if we choose h_1 such that $h_1 = h^{1/3}$, then the conditions of Lemma 6.1 hold.

Using an example we compare the quality of the modeling for the two means of approximation under consideration. Suppose that when integrating the initial system by a method of the second order of accuracy we decide to choose at $h = 0.001$. Then in the construction of the necessary random variables using the first method, $h_1 = h$ and we need $\approx 1/h$, i.e. ≈ 1000 normally distributed random variables. For the second method $h_1 = h^{1/3} = 0.1$; however, at each step we have to model not one but two random variables, $\Delta_k v(h_1)$ and $\int_{s_k}^{s_{k+1}} (v(s) - v(s_k))\,ds$. As a result, instead of 1000 random variables here we need only 20 variables. Of course, the second method is far more economic.

We now give a Lemma related with approximate modeling by the method (6.58).

LEMMA 6.3. *Suppose we propose to solve the system* (6.1) *by a method of second order of accuracy with step h in accordance with Theorem* 2.1, *in which at each step the integrals* $\int_{t_k}^{t_{k+1}} dw(\theta)$, $\int_{t_k}^{t_{k+1}} (w(\theta)-w(t_k))\,d\theta$, $\int_{t_k}^{t_{k+1}} (w(\theta)-w(t_k))^2\,d\theta$ *participate. If these random variables are replaced, independently of the step at which we are, by random variables* $h^{1/2}x_N$, $h^{3/2}y_N$, $h^2 z_N$, *where* x_N, y_N, z_N *can be found recurrently from* (6.58) *with step* $h_1 = O(h^{1/3})$, *then the order of accuracy of the initial method (which is* 2) *remains the same.*

For completeness of exposition we give a proof of the following useful relations:

$$\mathbf{E}\,(z(1)-\overline{z}(1)) = \mathbf{E}\left(\int\limits_{t_k}^{t_{k+1}} (w(\theta)-w(t_k))^2\,d\theta - z_N\right) = 0, \tag{6.59}$$

$$\mathbf{E}\,(z(1)-\overline{z}(1))^2 = \mathbf{E}\left(\int\limits_{t_k}^{t_{k+1}} (w(\theta)-w(t_k))^2\,d\theta - z_N\right)^2 = \frac{h_1^3}{3}. \tag{6.60}$$

We have (since $x_k = v(s_k)$, $Nh_1 = 1$):

$$\begin{aligned} z(1)-\overline{z}(1) &= \int\limits_0^1 v^2(s)\,ds - z_N \\ &= \sum_{k=0}^{N-1}\int\limits_{s_k}^{s_{k+1}} v^2(s)\,ds - \sum_{k=0}^{N-1} v^2(s_k)h_1 \\ &\quad - 2\sum_{k=0}^{N-1} v(s_k)\int\limits_{s_k}^{s_{k+1}} (v(s)-v(s_k))\,ds - \sum_{k=0}^{N-1}\frac{h_1^2}{2} \\ &= \sum_{k=0}^{N-1}\int\limits_{s_k}^{s_{k+1}} (v(s)-v(s_k))^2\,ds - \frac{h_1}{2}. \end{aligned}$$

This immediately implies (6.59). Further,

$$\begin{aligned} \mathbf{E}\,(z(1)-\overline{z}(1))^2 &= \sum_{k=0}^{N-1}\mathbf{E}\left(\int\limits_{s_k}^{s_{k+1}} (v(s)-v(s_k))^2\,ds\right)^2 \\ &\quad + 2\sum_{i=0}^{N-2}\sum_{k=i+1}^{N-1}\mathbf{E}\int\limits_{s_k}^{s_{k+1}} (v(s)-v(s_i))^2\,ds\cdot\mathbf{E}\int\limits_{s_k}^{s_{k+1}} (v(s)-v(s_k))^2\,ds - \frac{h_1^2}{4} \\ &= \sum_{k=0}^{N-1}\mathbf{E}\left(\int\limits_{s_k}^{s_{k+1}} (v(s)-v(s_k))^2\,ds\right)^2 - \frac{h_1^3}{4}. \end{aligned} \tag{6.61}$$

We compute $\mathbf{E}\left(\int_0^t v^2(s)\,ds\right)^2$. Put $Z(t) = \int_0^t v^2(s)\,ds$. We have:

$$dz^2(t) = 2zv^2\,dt, \qquad d(zv^2) = v^4\,dt + 2zv\,dt + z\,dt, \qquad dz = v^2\,dt.$$

Hence,

$$\begin{aligned} d\mathbf{E}z^2 &= 2\mathbf{E}(zv^2)\,dt, \qquad \mathbf{E}z^2\Big|_{t=0} = 0, \\ d\mathbf{E}(zv^2) &= \mathbf{E}v^4\,dt + \mathbf{E}z\,dt, \qquad \mathbf{E}(zv^2)\Big|_{t=0} = 0, \\ d\mathbf{E}z &= \mathbf{E}v^2\,dt, \qquad \mathbf{E}z\big|_{t=0} = 0. \end{aligned} \tag{6.62}$$

Since $\mathbf{E}v^2(t) = t$, $\mathbf{E}v^4(t) = 3t^2$, by successively solving the equations (6.62) from bottom to top we find

$$\mathbf{E}z^2(t) = \mathbf{E}\left(\int_0^t v^2(s)\,ds\right)^2 = \frac{7}{12}t^4. \tag{6.63}$$

Using (6.63) we obtain (6.60) from (6.61).

REMARK 6.1. Application of the usual numerical integration formulas for modeling the integrals does not lead to success. We will convince ourselves of this by the example of modeling the integral $\int_0^1 v(s)\,ds$ using the trapezium formula. It is well known that the trapezium formula has error $O(h^2)$. This is true for integrands having bounded second derivative. Here, however, $v(s)$ is a Wiener process with nonsmooth trajectories. Of course we are interested in accuracy in the mean sense. More precisely, we are interested in mean and mean-square deviation of the approximation of the integral by the trapezium formula.

Applying the trapezium formula to the integral $\int_0^1 v(s)\,ds$ gives

$$\int_0^1 v(s)\,ds \doteq \frac{h}{2}\left(v(0) + 2v(s_1) + \cdots + 2v(s_{N-1}) + v(s_N)\right), \qquad h = \frac{1}{N}. \tag{6.64}$$

Here, the mean deviation is zero, since the mathematical expectations of both sides of (6.64) are zero. It is easy to compute that

$$\begin{aligned} &\mathbf{E}\left(\int_0^1 v(s)\,ds - \sum_{k=0}^{N-1}\frac{v(s_k)+v(s_{k+1})}{2}h\right)^2 \\ &= \sum_{k=0}^{N-1}\mathbf{E}\left(\int_{s_k}^{s_{k+1}}\left(v(s) - v(s_k) - \frac{v(s_{k+1})-v(s_k)}{2}\right)ds\right)^2 = \frac{h^2}{12}. \end{aligned}$$

Thus, the mean-square deviation in the trapezium method is $O(h)$, which is of lower order than expected.

7. Modeling Itô integrals depending on several noises

In the case of several noises the difficulty in modeling increases sharply. Of course, as in the case of a single noise, many relations exist between the Itô integrals appearing in some method or other, and this allows us to reduce the number of random variables to be modeled. But there are considerably more such variables, they have a more complicated structure, and they are interdependent. Even for a method of the first order, in general there is clearly no real possibility of modeling exactly the required random variables. Therefore here it is highly necessary to model approximately.

7.1. Exact methods for modeling the random variables in a method of order 1 in the case of two noises. For a method of the first order of accuracy we have to model $w_1(h)$, $w_2(h)$, and $\int_0^h w_2(s)\,dw_1(s)$, since $\int_0^h w_1(s)\,dw_2(s) = w_1(h)w_h(h) - \int_0^h w_2(s)\,dw_1(s)$. To model these three random variables we consider the system of equations

$$\begin{aligned} d\eta_1 &= dw_1, & \eta_1(0) &= 0, \\ d\eta_2 &= \xi dw_1, & \eta_2(0) &= 0, \\ d\xi &= dw_2, & \xi(0) &= 0. \end{aligned} \tag{7.1}$$

It is well known (see [23, p. 472]) that the conditional distribution $\mathbf{P}(\eta_1 < a_1,\ \eta_2 < a_2 \mid \xi(s),\ 0 \le s \le t)$ is Gaussian. The parameters m_t and γ_t of this distribution can be readily computed (see [23]):

$$m_t = \mathbf{E}\left(\eta(t) \mid \xi(s),\ 0 \le s \le t\right) = 0,$$

$$\gamma_t = \mathbf{E}\left(\left(\eta(t) - m_t\right)\left(\eta(t) - m_t\right)^{\mathrm{T}} \mid \xi(s),\ 0 \le s \le t\right) = \begin{bmatrix} t & \int_0^t \xi(s)\,ds \\ \int_0^t \xi(s)\,ds & \int_0^t \xi^2(s)\,ds \end{bmatrix}, \tag{7.2}$$

where $\eta(t)$ is the two-dimensional vector with components $\eta_1(t)$ and $\eta_2(t)$.

It can be seen that the parameters of the conditional distribution depend on $\xi(s)$ only by means of $\int_0^t \xi(s)\,ds = \int_0^t w_2(s)\,ds$ and $\int_0^t w_2^2(s)\,ds$. This implies the following rule for modeling $w_1(h)$, $w_2(h)$, and $\int_0^h w_2(s)\,dw_1(s)$. First we model $w_2(h)$, $\int_0^h w_2(s)\,ds$, and $\int_0^h w_2^2(s)\,ds$. Then we model pairs of random variables having the normal distribution with characteristics

$$m_h = 0, \qquad \gamma_h = \begin{bmatrix} h & \int_0^h w_2(s)\,ds \\ \int_0^h w_2(s)\,ds & \int_0^h w_2^2(s)\,ds \end{bmatrix}. \tag{7.3}$$

To conveniently model such pairs of random variables, having obtained $w_1(h)$ by the law $N(0,h)$, we use the following representation:

$$\int_0^h w_2(s)\,dw_1(s) = \frac{1}{h} w_1(h) \int_0^h w_2(s)\,ds + \left(\int_0^h w_2(s)\,dw_1(s) - \frac{1}{h} w_1(h) \int_0^h w_2(s)\,ds \right). \tag{7.4}$$

It can be readily seen that the two random variables appearing as terms in this sum are conditionally Gaussian distributed (under the condition that $w_2(h)$, $\int_0^h w_2(s)\,ds$, and $\int_0^h w_2^2(s)\,ds$ are known) and are conditionally independent. Moreover, the second term has mathematical expectation zero and variance $\int_0^h w_2^2(s)\,ds - \left(\int_0^h w_2(s)\,ds\right)^2/h$. Thus, the main problem in constructing the initial variables consists of modeling $w_2(h)$, $\int_0^h w_2(s)\,ds$, and $\int_0^h w_2^2(s)\,ds$. Earlier we have already convinced ourselves of the fact that this is a rather serious problem. It is clear that as the amount of noises increases, the difficulty of exact modeling grows in an unmeasurable manner.

7.2. Use of the numerical integration of special linear stochastic systems for modeling Itô integrals. The idea of this manner of modeling does not differ from the one expounded in Subsection 6.4. We first of all show that all Itô integrals up to half-integral order m inclusively satisfy a linear (autonomous) system of stochastic differential equations. In fact, all Itô integrals of order 1/2 satisfy the system of equations

$$dI_1 = dw_1, \dots, dI_q = dw_q. \tag{7.5}$$

Further, all Itô integrals of order 1 satisfy the system of equations

$$dI_{ij} = I_i\,dw_j, \qquad i \neq 0, \quad j \neq 0, \qquad dI_0 = dt. \tag{7.6}$$

Having a system of equations for all Itô integrals up to order m inclusively (here and below we take 'all integrals' to mean the integrals participating in Taylor-type expansions), we can obtain a system for all Itô integrals of order $m+1/2$ by adhering to the following considerations. Consider all Itô integrals of orders $m-1/2$ and m. If $I_{i_1,\dots,i_k}(\theta)$ is an Itô integral of order $m-1/2$, then $I_{i_1,\dots,i_k,i_{k+1}}$, where $i_{k+1}=0$, is an Itô integral of order $m+1/2$, and it satisfies the equation

$$dI_{i_1,\dots,i_k,i_{k+1}} = I_{i_1,\dots,i_k}\,d\theta. \tag{7.7}$$

If $I_{i_1,\dots,i_l}$ is an Itô integral of order m, then $I_{i_1,\dots,i_l,i_{l+1}}$, $i_{l+1}=1,\dots,q$, satisfies the equation

$$dI_{i_1,\dots,i_l,i_{l+1}} = I_{i_1,\dots,i_l}\,dw_{i_{l+1}}(\theta). \tag{7.8}$$

As a result we have accounted for all Itô integrals of order $m+1/2$. Adjoining the equations of the type (7.7) and (7.8) to the system of equations for the Itô integrals up to order m, we obtain a system of equations for the Itô integrals up to order $m+1/2$ inclusively.

It can be readily seen that this system is a linear autonomous system of stochastic differential equations. The initial data for each variable is zero. It is clear that the dimension of this system can be substantially reduced because of relations between the integrals.

As in Subsection 6.4, here also it is convenient to pass to integration over, say, the interval $[0,1]$. It is then necessary to make the change of variables

$$s = \frac{\theta}{h}, \qquad v_i(s) = \frac{w_i(sh)}{\sqrt{h}}, \qquad 0 \le s \le 1, \qquad i = 1,\dots,q. \tag{7.9}$$

After this, the integral $I_{i_1,\dots,i_k}$ becomes

$$I_{i_1,\dots,i_k}(h;w) = h^{\sum_{j=1}^{k}(2-\bar{\imath}_j)/2} \int_0^1 dv_{i_k}(s) \int_0^s dv_{i_{k-1}}(s_1) \cdots \int_0^{s_{k-2}} dv_{i_1}(s_{k-1}), \tag{7.10}$$

and the problem reduces to modeling the integrals

$$I_{i_1,\dots,i_k}(t;v) = \int_0^t dv_{i_k}(s) \int_0^s \cdots \int_0^{s_{k-2}} dv_{i_1}(s_{k-1}) \tag{7.11}$$

for $t = 1$. The integrals (7.11) satisfy the same system of equations as the integrals (7.10), but because the orders of the various integrals are different, the integrals (7.11) can (and must) be modeled with different orders of accuracy. To clarify the order of accuracy of the modeling we can be guided by Lemma 6.2.

7.3. Modeling the Itô integrals $\int_0^h w_i(s)\,dw_j(s)$, $i,j = 1,\dots,q$. A method of first order of accuracy for the system (1.1) has the form

$$\begin{gathered} X_0 = X(t_0), \\ X_{k+1} = X_k + \sum_{j=1}^{q} (\sigma_j)_k \Delta_k w_j(h) + a_k h \\ + \sum_{i=1}^{q}\sum_{j=1}^{q} (\Lambda_i \sigma_j)_k \int_{t_k}^{t_{k+1}} (w_i(s) - w_i(t_k))\,dw_j(s) \\ k = 0,\dots,N-1, \end{gathered} \tag{7.12}$$

where $t_0 < t_1 < \cdots < t_N = t_0 + T$, $h = t_{k+1} - t_k = T/N$. To realise the method (7.12), at each step we have to model the set of random variables $\Delta_k w_j$, $\int_{t_k}^{t_{k+1}} (w_i(s) - w_i(t_k))\,dw_j(s)$, $i,j = 1,\dots,q$. Since at different steps these sets are independent, the problem reduces to modeling the variables $w_j(h)$, $I_{ij} = \int_0^h w_i(s)\,dw_j(s)$, $i,j = 1,\dots,q$. Since

$$\int_0^h w_j(s)\,dw_i(s) = w_i(h)w_j(h) - \int_0^h w_i(s)\,dw_j(s), \qquad i \neq j, \tag{7.13}$$

$$\int_0^h w_i(s)\,dw_i(s) = \frac{w_i^2(h)}{2} - \frac{h}{2}, \tag{7.14}$$

it suffices to model the set of variables

$$w_j(h), \qquad j = 1,\dots,q,$$

$$I_{ij} = \int_0^h w_i(s)\,dw_j(s), \qquad i = 1,\dots,q, \quad j = i+1,\dots,q. \tag{7.15}$$

Consider the *rectangle method*. We write the integral $\int_0^h w_i(s)\,dw_j(s)$ as a sum of l integrals:

$$\int_0^h w_i(s)\,dw_j(s) = \sum_{k=1}^{l} \int_{s_{k-1}}^{s_k} w_i(s)\,dw_j(s),$$

$$s_0 = 0, \qquad s_k - s_{k-1} = \frac{h}{l}, \qquad k = 1,\dots,l. \tag{7.16}$$

We replace each of these integrals using the left rectangle formula and obtain

$$\int_0^h w_i(s)\,dw_j(s) \stackrel{\bullet}{=} \sum_{k=1}^{l} w_i(s_{k-1})\left(w_j(s_k) - w_j(s_{k-1})\right). \tag{7.17}$$

For the error

$$\Delta_{ij} = \sum_{k=1}^{l} \int_{s_{k-1}}^{s_k} \left(w_i(s) - w_i(s_{k-1})\right) dw_j(s)$$

in the approximate identity (7.17) we have

$$\mathbf{E}\Delta_{ij} = 1,$$

$$\begin{aligned} \mathbf{E}\Delta_{ij}^2 &= \sum_{k=1}^{l} \mathbf{E}\left(\int_{s_{k-1}}^{s_k} \left(w_i(s) - w_i(s_{k-1})\right) dw_j(s)\right)^2 \\ &= \sum_{k=1}^{l} \int_{s_{k-1}}^{s_k} \mathbf{E}\left(w_i(s) - w_i(s_{k-1})\right)^2 ds = \frac{1}{2}\frac{h^2}{l}. \end{aligned} \tag{7.18}$$

Thus, according to (7.17), to approximately represent the variables (7.15) we have to model ql independent $N(0,1)$-distributed random variables ξ_{ik}, $i = 1,\dots,q$, $k = 1,\dots,l$ ($\sqrt{h/l}\,\xi_{ik} = w_i(s_k) - w_i(s_{k-1})$) and put

$$w_i(h) = \sqrt{\frac{h}{l}} \sum_{r=1}^{l} \xi_{ir},$$

$$\int_0^h w_i(s)\,dw_j(s) \stackrel{\bullet}{=} \frac{h}{l} \sum_{k=1}^{l} \sum_{r=1}^{k-1} \xi_{ir}\xi_{jk},$$

$$i = 1,\dots,q, \qquad j = i+1,\dots,q. \tag{7.19}$$

Note that if the integral $\int_0^h w_i(s)\,dw_i(s)$ is modeled according to (7.19), then the error involved is $h^2/(2l)$, while this integral can be computed exactly by (7.14) at the same time. Further, the integrals for $i > j$ can be approximately computed either by using (7.13) after having approximately modeled (7.15), or by modeling them according to (7.17). In both cases the error will be the same. It is easy to see that the use of (7.13) is equivalent to the use of the right rectangle formula.

We now consider the *trapezium method*. Applying to each integral in the sum (7.16) the trapezium formula, we find

$$\int_0^h w_i(s)\,dw_j(s) \doteq \sum_{k=1}^{l} \frac{1}{2}\left(w_i(s_{k-1}) + w_i(s_k)\right)\left(w_j(s_k) - w_j(s_{k-1})\right). \tag{7.20}$$

For the error

$$\Delta_{ij} = \frac{1}{2}\sum_{k=1}^{l}\int_{s_{k-1}}^{s_k} \left((w_i(s) - w_i(s_{k-1})) - (w_i(s_k) - w_i(s))\right) dw_j(s)$$

in the approximate identity (7.20) we have

$$\mathbf{E}\Delta_{ij} = 0,$$

$$\begin{aligned}\mathbf{E}\Delta_{ij}^2 &= \frac{1}{4}\sum_{k=1}^{l}\int_{s_{k-1}}^{s_k} \mathbf{E}\left((w_i(s) - w_i(s_{k-1})) - (w_i(s_k) - w_i(s))\right)^2 ds \\ &= \frac{1}{4}\sum_{k=1}^{l}\left((s - s_{k-1}) + (s_k - s)\right) ds = \frac{1}{4}\frac{h^2}{l}.\end{aligned} \tag{7.21}$$

Here, having modeled $q \cdot l$ independent $N(0,1)$-distributed random variables ξ_{ik} we can set

$$w_i(h) = \sqrt{\frac{h}{l}}\sum_{k=1}^{l}\xi_{ir},$$

$$\int_0^h w_i(s)\,dw_j(s) \doteq \frac{h}{l}\sum_{k=1}^{l}\left(\sum_{r=1}^{k-1}\xi_{ir} + \frac{1}{2}\xi_{ik}\right)\xi_{jk}, \qquad i < j. \tag{7.22}$$

It can be readily seen that (7.13) gives the same result as (7.22) for $i > j$. If we take $l \approx 1/h$, then (7.18), (7.21) imply that the mean-square error of the approximations of the integrals is $O(h^{3/2})$ in both the rectangle and the trapezium method, while the mean error is zero. By Lemma 6.2 this implies the following result.

LEMMA 7.1. *If we replace in the method* (7.12) *at each step* (*irrespective at which step we are*) *the random variables involved using either the rectangle or the trapezium formula with* $l \approx 1/h$, *then the order of accuracy of the method does not become smaller.*

With respect to labour involved, both methods are identical; however, with respect to accuracy the trapezium method somewhat exceeds the rectangle method.

We now turn to a method which can be naturally called the *Fourier method*.

Consider the Fourier coefficients of the process $w_i(t) - (t/h)w_i(h)$ on the interval $0 \le t \le h$ with respect to the trigonometric system of functions 1, $\cos 2k\pi t/h$, $\sin 2k\pi t/h$,

$k = 1, 2, \ldots$ (see [5], [18] for the *Wiener construction* of Brownian motion). We have

$$a_{ik} = \frac{2}{h} \int_0^h \left(w_i(s) - \frac{s}{h} w_i(h) \right) \cos \frac{2\pi k s}{h} \, ds, \qquad k = 0, 1, 2, \ldots,$$

$$b_{ik} = \frac{2}{h} \int_0^h \left(w_i(s) - \frac{s}{h} w_i(h) \right) \sin \frac{2\pi k s}{h} \, ds, \qquad k = 1, 2, \ldots.$$

The distribution of these coefficients is clearly Gaussian.

LEMMA 7.2. *The following relations hold:*

$$\mathbf{E} w_i(h) a_{ik} = 0, \quad k = 0, 1, 2, \ldots, \qquad \mathbf{E} w_i(h) b_{ik} = 0, \quad k = 1, 2, \ldots, \tag{7.23}$$

$$\mathbf{E} a_{i0}^2 = \frac{h}{3}, \qquad \mathbf{E} a_{ik}^2 = \mathbf{E} b_{ik}^2 = \frac{h}{2k^2\pi^2}, \quad k = 1, 2, \ldots, \tag{7.24}$$

$$\mathbf{E} a_{i0} a_{ik} = -\frac{h}{k^2 \pi^2}, \qquad k = 1, 2, \ldots, \tag{7.25}$$

$$\mathbf{E} a_{i0} b_{ik} = \mathbf{E} a_{ik} a_{im} = \mathbf{E} a_{ik} b_{im} = \mathbf{E} b_{ik} b_{im} = 0, \qquad k, m = 1, 2, \ldots, \quad k \neq m. \tag{7.26}$$

PROOF. All these formulas can be obtained by direct computation. We give, as an example, a detailed proof of one of the formulas in (7.24). For $k \neq 0$:

$$\mathbf{E} a_{ik}^2 = \frac{4}{h^2} \int_0^h \int_0^h \mathbf{E} \left(\left(w(t) - \frac{t}{h} w(h) \right) \left(w(s) - \frac{s}{h} w(h) \right) \right) \cos \frac{2k\pi t}{h} \cos \frac{2k\pi s}{h} \, dt \, ds.$$

By computation,

$$\mathbf{E} \left(\left(w(t) - \frac{t}{h} w(h) \right) \left(w(s) - \frac{s}{h} w(h) \right) \right) = \begin{cases} t - \frac{ts}{h}, & t \le s, \\ s - \frac{ts}{h}, & t > s. \end{cases}$$

Therefore

$$\mathbf{E} a_{ik}^2 = \frac{4}{h^2} \int_0^h \left(\int_0^s t \cos \frac{2k\pi t}{h} \, dt \right) \cos \frac{2k\pi s}{h} \, ds$$
$$+ \frac{4}{h^2} \int_0^h \left(\int_s^h \cos \frac{2k\pi t}{h} \, dt \right) s \cos \frac{2k\pi s}{h} \, ds - \frac{4}{h^2} \int_0^h \int_0^h \frac{st}{h} \cos \frac{2k\pi t}{h} \cos \frac{2k\pi s}{h} \, dt \, ds. \tag{7.27}$$

Further,

$$\int_0^s t \cos \frac{2k\pi t}{h} \, dt = s \sin \frac{2k\pi s}{h} \cdot \frac{h}{2k\pi} + \left(\frac{h}{2k\pi} \right)^2 \left(\cos \frac{2k\pi s}{h} - 1 \right).$$

Hence, in particular,

$$\int_0^h t \cos \frac{2k\pi t}{h} \, dt = 0.$$

Hence the last term in (7.27) vanishes. Using these equations we obtain:

$$\begin{aligned}\mathbf{E}a_{ik}^2 &= \frac{4}{h^2}\int_0^h \left(s\sin\frac{2k\pi s}{h}\cdot\frac{h}{2k\pi} + \left(\frac{h}{2k\pi}\right)^2\left(\cos\frac{2k\pi s}{h}-1\right)\right)\cos\frac{2k\pi s}{h}\,ds \\ &+ \frac{4}{h^2}\int_0^h \left(-\frac{h}{2k\pi}\sin\frac{2k\pi s}{h}\right) s\cos\frac{2k\pi s}{h}\,ds \\ &= \frac{4}{h^2}\int_0^h \left(\frac{h}{2k\pi}\right)^2\left(\cos\frac{2k\pi s}{h}-1\right)\cos\frac{2k\pi s}{h}\,ds = \frac{h}{2k^2\pi^2}.\end{aligned}$$

The other formulas can be similarly proved. □

We replace the integrand $w_i(s)$ in the integral $I_{ij} = \int_0^h w_i(s)\,dw_j(s)$ by an expression containing a segment of its Fourier series:

$$\begin{aligned}I_{ij} &= \int_0^h w_i(s)\,dw_j(s) \\ &\doteq \overline{I}_{ij} \\ &= \int_0^h \left(\frac{t}{h}w_i(h) + \frac{a_{i0}}{2} + \sum_{k=1}^m \left(a_{ik}\cos\frac{2k\pi t}{h} + b_{ik}\sin\frac{2k\pi t}{h}\right)\right) dw_j(t).\end{aligned} \tag{7.28}$$

LEMMA 7.3. *For the error Δ_{ij} in the approximate identity* (7.28) *the following relation holds:*

$$\mathbf{E}\Delta_{ij}^2 = \frac{h^2}{12} - \frac{h^2}{2\pi^2}\sum_{k=1}^m \frac{1}{k^2}, \qquad i \neq j. \tag{7.29}$$

PROOF. For $i \neq j$,

$$\begin{aligned}\mathbf{E}\Delta_{ij}^2 &= \mathbf{E}\left(\int_0^h \left(w_i(t) - \frac{t}{h}w_i(h) - \frac{a_{i0}}{2} - \sum_{k=1}^m \left(a_{ik}\cos\frac{2k\pi t}{h} + b_{ik}\sin\frac{2k\pi t}{h}\right)\right) dw_j(t)\right)^2 \\ &= \mathbf{E}\int_0^h \left(w_i(t) - \frac{t}{h}w_i(h) - \frac{a_{i0}}{2} - \sum_{k=1}^m \left(a_{ik}\cos\frac{2k\pi t}{h} + b_{ik}\sin\frac{2k\pi t}{h}\right)\right)^2 dt.\end{aligned}$$

Since

$$\frac{a_{i0}}{2} + \sum_{k=1}^m a_{ik}\cos\frac{2k\pi t}{h} + b_{ik}\sin\frac{2k\pi t}{h}$$

is a segment of the Fourier series of the function $w_i(t) - (t/h)w_i(h)$, we have

$$\int_0^h \left(w_i(t) - \frac{t}{h}w_i(h) - \frac{a_{i0}}{2} - \sum_{k=1}^m \left(a_{ik} \cos\frac{2k\pi t}{h} + b_{ik} \sin\frac{2k\pi t}{h} \right) \right)^2 dt$$
$$= \int_0^h \left(w_i(t) - \frac{t}{h}w_i(h) \right)^2 dt - \frac{1}{4}ha_{i0}^2 - \sum_{k=1}^m \frac{h}{2} \left(a_{ik}^2 + b_{ik}^2 \right).$$

Using Lemma 7.2 we thus find

$$\mathbf{E}\Delta_{ij}^2 = \int_0^h \mathbf{E} \left(w_i(t) - \frac{t}{h}w_i(h) \right)^2 dt - \frac{h}{4}\mathbf{E}a_{i0}^2 - \sum_{k=1}^m \frac{h}{2}\mathbf{E} \left(a_{ik}^2 + b_{ik}^2 \right)$$
$$= \int_0^h \left(t - \frac{2t^2}{h} + \frac{t^2}{h} \right) dt - \frac{1}{12}h^2 - \sum_{k=1}^m \frac{h}{2} \cdot \frac{h}{k^2\pi^2}.$$

This implies (7.29). □

LEMMA 7.4. *The following formula is true:*

$$\bar{I}_{ij} = \int_0^h \left(\frac{t}{h}w_i(h) + \frac{a_{i0}}{2} + \sum_{k=1}^m \left(a_{ik} \cos\frac{2k\pi t}{h} + b_{ik} \sin\frac{2k\pi t}{h} \right) \right) dw_j(t)$$
$$= \frac{1}{2}w_i(h)w_j(h) + \frac{a_{i0}}{2}w_j(h) - \frac{a_{j0}}{2}w_i(h) + \pi \sum_{k=1}^m k \left(a_{ik}b_{jk} - b_{ik}a_{jk} \right). \tag{7.30}$$

PROOF. We have

$$\int_0^h t\, dw_j(t) = hw_j(h) - \int_0^h w_j(t)\, dt,$$
$$\int_0^h \cos\frac{2k\pi t}{h}\, dw_j(t) = w_j(h) + \frac{2k\pi}{h} \int_0^h w_j(t) \sin\frac{2k\pi t}{h}\, dt,$$
$$\int_0^h \sin\frac{2k\pi t}{h}\, dw_j(t) = -\frac{2k\pi}{h} \int_0^h w_j(t) \cos\frac{2k\pi t}{h}\, dt. \tag{7.31}$$

Further (cf. the formulas defining the Fourier coefficients):

$$a_{j0} = \frac{2}{h}\int_0^h w_j(t)\,dt - w_j(h),$$

$$a_{jk} = \frac{2}{h}\int_0^h \left(w_j(t) - \frac{t}{h}w_j(h)\right)\cos\frac{2k\pi t}{h}\,dt$$

$$= \frac{2}{h}\int_0^h w_j(t)\cos\frac{2k\pi t}{h}\,dt, \qquad k = 1,2,\ldots,$$

$$b_{jk} = \frac{2}{h}\int_0^h \left(w_j(t) - \frac{t}{h}w_j(h)\right)\sin\frac{2k\pi t}{h}\,dt. \tag{7.32}$$

Transforming first the expression for $\overline{I}_{ij}$ by using (7.31), and then by using (7.32), after a few computations we are led to (7.30). In these computations we use the identities

$$\int_0^h t\cos\frac{2k\pi t}{h}\,dt = 0, \qquad \int_0^h t\sin\frac{2k\pi t}{h}\,dt = -\frac{h^2}{2k\pi}.$$

This proves the Lemma. □

At the righthand side of (7.30) the coefficient a_{i0} depends on a_{ik}, the coefficient a_{j0} depends on a_{jk}, and all remaining coefficients (see Lemma 7.2) are mutually independent. Introduce the new random variable $a_{i0}^{(m)}$ by

$$a_{i0}^{(m)} = -\frac{a_{i0}}{2} - \sum_{k=1}^{m} a_{ik}. \tag{7.33}$$

We show that $a_{i0}^{(m)}$ does not depend on a_{ik}, $k = 1,\ldots,m$. In fact, by Lemma 7.2 we have

$$\mathbf{E}a_{i0}^{(m)}a_{ik} = -\frac{1}{2}\mathbf{E}a_{i0}a_{ik} - \mathbf{E}a_{ik}^2 = 0, \qquad k = 1,\ldots,m.$$

In view of the fact that all variables under consideration are Gaussian, this implies independence.

We can immediately compute that

$$\mathbf{E}\left(a_{i0}^{(m)}\right)^2 = \frac{h}{12} - \frac{h}{2\pi^2}\sum_{k=1}^{m}\frac{1}{k^2}. \tag{7.34}$$

Substituting $a_{i0} = -2a_{i0}^{(m)} - 2\sum_{k=1}^{m} a_{ik}$ into (7.30), we obtain

$$\overline{I}_{ij} = \frac{1}{2}w_i(h)w_j(h) + a_{j0}^{(m)}w_i(h) - a_{i0}^{(m)}w_j(h)$$
$$+ \sum_{k=1}^{m}\left(a_{jk}w_i(h) - a_{ik}w_j(h)\right) + \pi\sum_{k=1}^{m} k\left(a_{ik}b_{jk} - b_{ik}a_{jk}\right). \tag{7.35}$$

In (7.35) all $w_i(h)$, $w_j(h)$, $a_{i0}^{(m)}$, $a_{j0}^{(m)}$, a_{ik}, a_{jk}, b_{ik}, b_{jk} are independent Gaussian random variables.

We gather the results concerning the approximate modeling of the variables $w_i(h)$ and $I_{ij} = \int_0^h w_i(s)\,dw_j(s)$ obtained above in the following Theorem.

THEOREM 7.1. *Making up $\overline{I}_{ij}$, $i, j = 1, \dots, q$, reduces to modeling $2(m+1)q$ independent $N(0,1)$-distributed random variables ξ_i, ξ_{ik}, $k = 0, \dots, m$, η_{ik}, $k = 1, \dots, m$. Here,*

$$\xi_i = h^{-1/2} w_i(h), \qquad \xi_{i0} = \left(\frac{h}{12} - \frac{h}{2\pi^2}\sum_{k=1}^m \frac{1}{k^2}\right)^{-1/2} a_{i0}^{(m)},$$
$$\xi_{ik} = \sqrt{2}k\pi h^{-1/2} a_{ik}, \qquad \eta_{ik} = \sqrt{2}k\pi h^{-1/2} b_{ik}.$$

The quantities $w_i(h)$ and $I_{ij}(h)$ can be expressed in terms of these variables by means of the formulas

$$w_i(h) = h^{1/2}\xi_i, \qquad I_{ii} = \overline{I}_{ii} = \frac{h}{2}(\xi_i^2 - 1), \qquad i = 1, \dots, q,$$
$$I_{ij} \doteq \overline{I}_{ij} = \frac{h}{2}\xi_i\xi_j + h\sqrt{\frac{1}{12} - \frac{1}{2\pi^2}\sum_{k=1}^m \frac{1}{k^2}}(\xi_{j0}\xi_i - \xi_{i0}\xi_j)$$
$$+ \frac{h}{\pi\sqrt{2}}\sum_{k=1}^m \frac{1}{k}(\xi_{jk}\xi_i - \xi_{ik}\xi_j) + \frac{h}{2\pi}\sum_{k=1}^m \frac{1}{k}(\xi_{ik}\xi_{jk} - \xi_{jk}\eta_{ik}), \tag{7.36}$$
$$i \neq j$$

The error $\Delta_{ij} = I_{ij} - \overline{I}_{ij}$ of the approximate modeling is characterised by the relations

$$\mathbf{E}\Delta_{ij} = 0,$$
$$\mathbf{E}\Delta_{ij}^2 = \frac{h^2}{12} - \frac{h^2}{2\pi^2}\sum_{k=1}^m \frac{1}{k^2}, \qquad i \neq j. \tag{7.37}$$

If $m \approx 1/h$, then the following method (which is constructive from the point of view of modeling random variables):

$$X_0 = X(t_0),$$
$$X_{k+1} = X_k + \sum_{j=1}^q (\sigma_j)_k \xi_i^{(k)} h^{1/2} + a_k h + \sum_{i=1}^q\sum_{j=1}^q (\Lambda_i\sigma_j)_k \overline{I}_{ij}^{(k)},$$
$$k = 0, \dots, N-1, \tag{7.38}$$

where the index k indicates that the random variables are modeled according to (7.36) *independently of at which step we are, is a method of the first order of accuracy for integrating* (1.1).

We only need prove the last assertion, since all previous ones follow from Lemma 7.2–Lemma 7.4. However, its proof does not differ at all from the proof of Lemma 7.1 if we take into account that (since $\sum_{k=1}^{\infty} 1/k^2 = \pi^2/6$)

$$\frac{1}{12} - \frac{1}{2\pi^2}\sum_{k=1}^{m}\frac{1}{k^2} = \frac{1}{2\pi^2}\left(\frac{\pi^2}{6} - \sum_{k=1}^{m}\frac{1}{k^2}\right) = \frac{1}{2\pi^2}\sum_{k=m+1}^{\infty}\frac{1}{k^2} \leq \frac{1}{2\pi^2}\int_m^{\infty}\frac{dx}{x^2} = \frac{1}{2\pi^2 m}.$$

We will now compare results of modeling by the rectangle method, the trapezium method, and the Fourier method. It is clear from (7.18) and (7.21) that to achieve the same accuracy, the rectangle method requires twice as many independent $N(0,1)$-distributed random variables as does the trapezium method. We compare the Fourier method and the trapezium method from this point of view. To this end we put $l = 2(m+1)$ in (7.21). Then for identical expenditure as regards forming random variables (since in the trapezium method we have to model $q \cdot l = 2(m+1)q$ independent $N(0,1)$-distributed random variables in order to approximately form all the necessary Itô integrals by (7.22)) the error in the trapezium method can be computed from (7.21), and that of the Fourier method from (7.37). We have given some values of the coefficients at h^2 in the error $\mathbf{E}\Delta_{ij}^2$ (Table 7.1): in the second row stand the coefficients in the trapezium method; in the third row those in the Fourier method. For m large the Fourier method is, for example, 2.5 times more economic than the trapezium method, since the error of the Fourier method is close to $1/(20\,m)$ while that of the trapezium method is close to $1/(8\,m)$.

Table 7.1. Coefficients of the error in the trapezium and the Fourier method.

m	1	2	3	4	5	10	20
$\frac{1}{8(m+1)}$	0.0625	0.0417	0.0312	0.0250	0.0208	0.0114	0.0060
$\frac{1}{12} - \frac{1}{2\pi^2}\sum_{k=1}^{m}\frac{1}{k^2}$	0.0327	0.0200	0.0144	0.0112	0.0092	0.0048	0.0025

CHAPTER 3

Weak approximation of solutions of systems of stochastic differential equations

As already mentioned in the Introduction, in cases when the modeling of solutions is intended for the application of Monte-Carlo methods we can refrain from mean-square approximations and use approximations that are in may respect simpler: weak approximations of solutions.

Recall that an approximate solution $\overline{X}(t)$ *approximates the solution* $X(t)$ *in the weak sense with (weak) order of accuracy* p (or $O(h^p)$) if the following inequality holds:

$$\left|\mathbf{E}f(\overline{X}(t)) - \mathbf{E}f(X(t))\right| = O(h^p), \qquad t_0 \le t \le t_0 + T,$$

for all f from a sufficiently large class of functions.

In this Chapter the word 'weak' will be omitted if this does not lead to misunderstanding.

8. One-step approximation

The *one-step weak approximation* $\overline{X}_{t,x}(t+h)$ of the solution $X_{t,x}(t+h)$ can be constructed from computing the nearness of moments (from the first up till the r-th inclusively) of the vector $\overline{X}_{t,x}(t+h) - x$ to the corresponding moments of the vector $X_{t,x}(t+h) - x$. The order of accuracy of the one-step approximation depends in this case on both the order of the moments under consideration and on the order of nearness of these.

To construct the one-step approximations of third order of accuracy considered in this Section, we have to take into account all moments up to order six inclusively.

8.1. Initial assumptions and notations. Lemmas on properties of remainders and Itô integrals. As before we consider the system

$$dX = a(t,X)\,dt + \sum_{r=1}^{q} \sigma_r(t,X)\,dw_r(t). \tag{8.1}$$

In (8.1), X, $a(t,X)$, and $\sigma_r(t,X)$ are vectors of dimension n with components X^i, a^i, σ_r^i. We assume that the functions $a(t,x)$, $\sigma_r(t,x)$ are sufficiently smooth with respect to the variables t, x and satisfy a Lipschitz condition with respect to x: for all $t \in [t_0, t_0 + T]$, $x \in \mathbb{R}^n$, $y \in \mathbb{R}^n$ the following inequality holds:

$$|a(t,x) - a(t,y)| + \sum_{r=1}^{q} |\sigma_r(t,x) - \sigma_r(t,y)| \leq K\,|x-y|\,. \tag{8.2}$$

Here and below $|x|$ denotes the euclidean norm of the vector x, and xy denotes the scalar product of the vectors x, y. We introduce operators

$$\Lambda_r f = \left(\sigma_r, \frac{\partial}{\partial x}\right) f,$$

$$Lf = \left(\frac{\partial}{\partial t} + \left(a, \frac{\partial}{\partial x}\right) + \frac{1}{2}\sum_{r=1}^{q}\sum_{i=1}^{n}\sum_{j=1}^{n} \sigma_r^i \sigma_r^j \frac{\partial^2}{\partial x^i \partial x^j}\right) f.$$

Here, f may be a scalar function or a vector-function.

In the course of exposition we will impose additional conditions on a and σ_r. Note that the conditions on a and σ_r given in Theorem 2.1 are sufficient for all results in this Section to hold. We recall that these conditions are related with the growth of functions of the form $\Lambda_{i_j} \dots \Lambda_{i_1} f(t,x)$ for $f \equiv x$ as $|x| \to \infty$ (see (2.24)); more precisely, these functions grow with respect to x at most as a a linear function of $|x|$ as $|x| \to \infty$. The indices $i_1, \dots, i_j$ take the values $0, 1, \dots, q$, and $\Lambda_0 = L$.

We rewrite (2.28):

$$\begin{aligned} X_{t,x}(t+h) = x &+ \sum_{r=1}^{q} \sigma_r \int_t^{t+h} dw_r(t) + ah + \sum_{r=1}^{q}\sum_{i=1}^{q} \Lambda_i \sigma_r \int_t^{t+h} (w_i(\theta) - w_i(t))\, dw_r(\theta) \\ &+ \sum_{r=1}^{q} L\sigma_r \int_t^{t+h} (\theta - t)\, dw_r(\theta) + \sum_{r=1}^{q} \Lambda_r a \int_t^{t+h} (w_r(\theta - w_r(t))\, d\theta \\ &+ \sum_{r=1}^{q}\sum_{i=1}^{q}\sum_{s=1}^{q} \Lambda_s \Lambda_i \sigma_r \int_t^{t+h} \left(\int_t^{\theta} (w_s(\theta_1) - w_s(\theta))\, dw_r(\theta_1)\right) dw_r(\theta) \\ &+ La\frac{h^2}{2} + \rho. \end{aligned} \tag{8.3}$$

In (8.3), the coefficients σ_r, a, $\Lambda_i\sigma_r$, $L\sigma_r$, $\Lambda_r a$, $\Lambda_s\Lambda_i\sigma_r$, La are to be computed at the point (t, x), while the remainder ρ can be computed by (2.29) (we do not write out ρ here).

DEFINITION 8.1. We say that a function $f(x)$ *belongs to the class* F, written as $f \in F$, if we can find constants $K > 0$, $\kappa > 0$ such that for all $x \in \mathbb{R}^n$ the following inequality holds:

$$|f(x)| \leq K\,(1 + |x|^{\kappa})\,. \tag{8.4}$$

If a function $f(s, x)$ depends not only on $x \in \mathbb{R}^n$ but also on a parameter $s \in S$, then we say that $f(s, x)$ *belongs to* F (*with respect to the variable* x) if an inequality of the type (8.4) holds uniformly in $s \in S$.

In the sequel we will need that σ_r, a, $\Lambda_i\sigma_r$, $L\sigma_r$, $\Lambda_r a$, La, $\Lambda_s\Lambda_i\sigma_r$, etc. belong to the class F. For example, in the proof of Lemma 8.1 (see below) we will use that fact that all integrands participating in the remainder ρ, as well as all functions obtained by applying the operators Λ_l, $l = 1, \dots, q$, and L to the functions $\Lambda_j\Lambda_s\Lambda_i\sigma_r$, $L\Lambda_i\sigma_r$, $\Lambda_i L\sigma_r$, $\Lambda_i\Lambda_r a$, belong to the class F. It is easy to see that for this to happen it suffices that all partial derivatives up to order five, inclusively, of the coefficients a, σ_r with respect to t and x belong to F. For brevity reasons, below we will not list all functions that are required to belong to F. In such cases we will assert that: the coefficients a, σ_r, $r = 1, \dots, q$, together with the partial derivatives of sufficiently high order belong to F.

LEMMA 8.1. *Suppose the Lipschitz condition* (8.2) *holds and suppose that the functions a, σ_r, $r = 1, \dots, q$, together with the partial derivatives of sufficiently high order belong to F. Then the following inequalities hold:*

$$|\mathbf{E}\rho| \le K(x)h^3, \qquad K(x) \in F, \tag{8.5}$$

$$\left|\mathbf{E}\rho^2\right| \le K(x)h^4, \qquad K(x) \in F, \tag{8.6}$$

$$\left|\mathbf{E}\rho \int_t^{t+h} dw_r(\theta)\right| \le K(x)h^3, \qquad K(x) \in F. \tag{8.7}$$

PROOF. The form of the remainder ρ (see (2.29)) and the fact that $L^2 a \in F$ imply that we can find an even number $2m$ and a number $K > 0$ such that

$$\begin{aligned} |\mathbf{E}\rho| &= \left|\mathbf{E}\int_t^{t+h}\left(\int_t^{\theta}\left(\int_t^{\theta_1} L^2 a(\theta_2, X(\theta_2))\, d\theta_2\right) d\theta_1\right) d\theta\right| \\ &\le \left|\int_t^{t+h}\left(\int_t^{\theta}\left(\int_t^{\theta_1} K\left(1 + \mathbf{E}\,|X(\theta_2)|^{2m}\right) d\theta_2\right) d\theta_1\right) d\theta\right|. \end{aligned} \tag{8.8}$$

But by [8, p. 48], $\mathbf{E}\,|X(\theta_2)|^{2m}$ is bounded by a quantity $K(1 + |x|^{2m})$. Hence (8.8) implies (8.5). In the proof of (8.6) we use the fact that each term (more precisely, the mathematical expectation of the norm of each term in (2.29)) is, in any case, of second order of smallness with respect to h. To prove (8.7) we have to treat each integral in the first four sums in (2.29) by Itô's formula. For example,

$$\int_t^{t+h}\left(\int_t^{\theta}\left(\int_t^{\theta_1}\left(\int_t^{\theta_2}\Lambda_j\Lambda_s\Lambda_i\sigma_r(\theta_3,X(\theta_3))\,dw_j(\theta_3)\right)dw_s(\theta_2)\right)dw_i(\theta_1)\right)dw_r(\theta)$$

$$=\Lambda_j\Lambda_s\Lambda_i\sigma_r(t,x)\int_t^{t+h}\left(\int_t^{\theta}\left(\int_t^{\theta_1}\left(\int_t^{\theta_2}dw_j(\theta_3)\right)dw_s(\theta_2)\right)dw_i(\theta_1)\right)dw_r(\theta)$$

$$+\sum_{l=1}^{q}\int_t^{t+h}\left(\int_t^{\theta}\left(\int_t^{\theta_1}\left(\int_t^{\theta_2}\left(\int_t^{\theta_3}\Lambda_l\Lambda_j\Lambda_s\Lambda_i\sigma_r(\theta_4,X(\theta_4))\,dw_l(\theta_4)\right)dw_j(\theta_3)\right)\right.\right.\right.$$

$$\left.\left.\left.\times\ dw_s(\theta_2)\right)dw_i(\theta_1)\right)dw_r(\theta)$$

$$+\int_t^{t+h}\left(\int_t^{\theta}\left(\int_t^{\theta_1}\left(\int_t^{\theta_2}\left(\int_t^{\theta_3}L\Lambda_j\Lambda_s\Lambda_i\sigma_r(\theta_4,X(\theta_4))\,d\theta_4\right)dw_j(\theta_3)\right)\right.\right.\right.$$

$$\left.\left.\left.\times\ dw_s(\theta_2)\right)dw_i(\theta_1)\right)dw_r(\theta).$$

As a result ρ can be written as a sum of terms of second, or higher, order of smallness with respect to h. Moreover, the terms of second order look like one of the integrals

$$I_{risj}=\int_t^{t+h}\left(\int_t^{\theta}\left(\int_t^{\theta_1}\left(\int_t^{\theta_2}dw_j(\theta_3)\right)dw_s(\theta_2)\right)dw_i(\theta_1)\right)dw_r(\theta),$$

$$I_{r0i}=\int_t^{t+h}\left(\int_t^{\theta}\left(\int_t^{\theta_1}dw_i(\theta_2)\right)d\theta_1\right)dw_r(\theta),$$

$$I_{ri0}=\int_t^{t+h}\left(\int_t^{\theta}\left(\int_t^{\theta_1}d\theta_2\right)dw_i(\theta_1)\right)dw_r(\theta),$$

$$I_{0ri}=\int_t^{t+h}\left(\int_t^{\theta}\left(\int_t^{\theta_1}dw_i(\theta_2)\right)dw_r(\theta_1)\right)d\theta,$$

respectively, with nonrandom coefficients $\Lambda_j\Lambda_s\Lambda_i\sigma_r$, $\Lambda_iL\sigma_r$, $L\Lambda_i\sigma_r$, $\Lambda_i\Lambda_r a$. It is easy to prove that the mathematical expectation of the product of $\int_t^{t+h}dw_r(\theta)$ with any term of second order of smallness is zero. For example, we can show that

$$\mathbf{E}\left(\int_t^{t+h}\left(\int_t^{\theta}\left(\int_t^{\theta_1}\left(\int_t^{\theta_2}dw_j(\theta_3)\right)dw_s(\theta_2)\right)dw_i(\theta_1)\right)dw_r(\theta)\cdot\int_t^{t+h}dw_l(\theta)\right)$$

$$=\mathbf{E}(I_{risj}\cdot I_l)=0. \tag{8.9}$$

In fact, by changing variables

$$v_k = -w_k, \qquad k = 1, \ldots, q,$$

the v_k will be independent Wiener processes. Since the Wiener processes participating in (8.9) are odd, we have

$$\mathbf{E}(I_{risj} \cdot I_l) = -\mathbf{E}\left(\int_t^{t+h}\left(\int_t^{\theta}\left(\int_t^{\theta_1}\left(\int_t^{\theta_2} dv_j(\theta_3)\right) dv_s(\theta_2)\right) dv_i(\theta_1)\right) dv_r(\theta) \cdot \int_t^{t+h} dv_l(\theta)\right)$$
$$= -\mathbf{E}(I_{risj} \cdot I_l),$$

which implies (8.9).

The other terms in ρ have order of smallness at least $5/2$. Using the Bunyakovsky–Schwarz inequality we can readily show that the mathematical expectation of the modulus of the product of each of such terms with $\int_t^{t+h} dw_l(\theta)$ is smaller than or equal to $K(x)h^3$, with $K(x) \in F$. This proves (8.7), and hence the Lemma. □

Using the identity

$$\int_t^{t+h} (\theta - t)\, dw_r(\theta) = h \int_t^{t+h} dw_r(\theta) - \int_t^{t+h} (w_r(\theta) - w_r(t))\, d\theta$$

and the notations

$$I_j = \int_t^{t+h} dw_j(\theta) = w_j(t+h) - w_j(t),$$

$$I_{jp} = \int_t^{t+h} (w_j(\theta) - w_j(t))\, dw_p(\theta),$$

$$I_{sir} = \int_t^{t+h}\left(\int_t^{\theta} (w_s(\theta_1) - w_s(t))\, dw_i(\theta_1)\right) dw_r(\theta),$$

$$J_r = \int_t^{t+h} (w_r(\theta) - w_r(t))\, d\theta,$$

we can introduce $\widetilde{X}$ by the formula

$$\widetilde{X} = x + \sum_{r=1}^{q} \sigma_r I_r + ah + \sum_{r=1}^{q}\sum_{i=1}^{q} \Lambda_i \sigma_r I_{ir}$$
$$+ \sum_{r=1}^{q} L\sigma_r \cdot I_r \cdot h + \sum_{r=1}^{q} (\Lambda_r a - L\sigma_r) \cdot J_r + La \cdot \frac{h^2}{2}. \tag{8.10}$$

We can rewrite (8.3) as follows:

$$X = \widetilde{X} + \sum_{r=1}^{q}\sum_{i=1}^{q}\sum_{s=1}^{q} \Lambda_s \Lambda_i \sigma_r \cdot I_{sir} + \rho. \tag{8.11}$$

LEMMA 8.2. *The following identities hold:*

$$\mathbf{E}I_{sir} = 0, \qquad \mathbf{E}I_{sir} \cdot I_j = 0, \qquad \mathbf{E}I_{sir} \cdot I_j \cdot I_p = 0,$$
$$\mathbf{E}I_{sir} \cdot I_{jp} = 0, \qquad i,j,p,r,s = 1,\dots,q. \tag{8.12}$$

PROOF. The first, third and fourth identities in (8.12) are obvious because of oddness considerations. We show the second identity. Without loss of generality we may put $t = 0$. For $j \neq s$, $j \neq i$, $j \neq r$ this identity follows from the independence of I_j and I_{sir}. To consider the other cases we introduce the system of equations

$$dx(\theta) = w_s(\theta)\, dw_i(\theta), \qquad x(0) = 0,$$
$$dy(\theta) = x(\theta)\, dw_r(\theta), \qquad y(0) = 0.$$

Then $I_{sir} = y(h)$ and $\mathbf{E}I_{sir} \cdot I_j = \mathbf{E}y(h) \cdot w_j(h)$. Let, e.g., $j = s$. Then by Itô's formula,

$$d\left(y(\theta)w_s(\theta)\right) = w_s(\theta)x(\theta)\, dw_r(\theta) + y(\theta)\, dw_s(\theta) + x(\theta)\delta_{sr}\, d\theta,$$

where δ_{sr} is the Kronecker symbol. Hence,

$$d\mathbf{E}\left(y(\theta)w_s(\theta)\right) = \mathbf{E}x(\theta)\delta_{sr}\, d\theta.$$

Since $\mathbf{E}x(\theta) = 0$ and $\mathbf{E}\left(y(0)w_s(0)\right) = 0$, we have $\mathbf{E}y(h)w_j(h) = \mathbf{E}y(h)w_s(h) = 0$. The cases $j = i$ and $j = r$ can be treated in a similar way. □

8.2. Forming one-step approximations of third order of accuracy. We introduce the notations $X = X(t+h)$, $\Delta = X - x$, $\widetilde{\Delta} = \widetilde{X} - x$, $\overline{\Delta} = \overline{X} - x$, and denote by x^i the ith coordinate of the vector x. Our nearest goal is to form a random vector $\overline{X}$ such that the difference of all moments up to order five, inclusively, of the coordinates of the vectors Δ and $\overline{\Delta}$ would have third order of smallness with respect to h. More precisely,

$$\left| \mathbf{E}\left(\prod_{j=1}^{s} \Delta^{i_j} - \prod_{j=1}^{s} \overline{\Delta}^{i_j} \right) \right| \le K(x)h^3, \qquad i_j = 1,\dots,n, \qquad s = 1,\dots,5, \qquad K(x) \in F, \tag{8.13}$$

and, moreover,

$$\mathbf{E}\prod_{j=1}^{s} \left|\overline{\Delta}^{i_j}\right| \le K(x)h^3, \qquad i_j = 1,\dots,n, \qquad s = 6, \qquad K(x) \in F, \tag{8.14}$$

First of all we state the following lemma.

LEMMA 8.3. *Under the conditions of Lemma* 8.1 *the following inequalities hold:*

$$\left| \mathbf{E}\left(\prod_{j=1}^{s} \Delta^{i_j} - \prod_{j=1}^{s} \widetilde{\Delta}^{i_j} \right) \right| \le K(x)h^3, \qquad s = 1,\dots,5, \qquad K(x) \in F. \tag{8.15}$$

PROOF. The proof of this Lemma is based on Lemma 8.1 and Lemma 8.2. In fact, by (8.11) each component Δ^{i_j} of Δ differs from the corresponding component $\widetilde{\Delta}^{i_j}$ of $\widetilde{\Delta}$ by a sum made up from the corresponding components of the vectors ρ_s and $\Lambda_s\Lambda_i\sigma_r \cdot I_{sir}$. Therefore the difference $\prod_{j=1}^{s} \Delta^{i_j} - \prod_{j=1}^{s} \widetilde{\Delta}^{i_j}$ consists of terms each of which must have as at least one of its factors a component of ρ_s or a component of the integral I_{sir}. If $s = 1$, i.e. we are considering first moments, then these terms do not have other factors, and (8.15) follows from (8.5) and the first identity in (8.12). If $s = 2$, then the terms containing ρ either have I_r as factor or they have a factor whose order is at least one. In the first case we use the estimate (8.7), and in the second case we use the Bunyakovsky–Schwarz inequality and, subsequently, (8.6). In other words, in the second case the order of a term is at least three because one factor has order at least one and the second (which is some component of ρ) has, by (8.6), order two, i.e. by the Bunyakovsky–Schwarz inequality we may add the orders of the factors. This already makes clear that for $s = 3, 4, 5$ the terms containing as factor at least one component of ρ have order of smallness at least three with respect to h. We return for $s = 2$ to the terms containing I_{sir} as factor; I_{sir} has order of smallness 3/2 with respect to h. Such terms either contain as factor an expression of the form $I_{sir} \cdot I_j$, $I_{sir} \cdot h$, $I_{sir} \cdot I_{jp}$, and the mathematical expectation of such terms is zero (by the first, second, and fourth identity in (8.12)), or, in any case, they have third order of smallness with respect to h. For $s = 3$, the terms containing I_{sir} and having a mean-square order of smallness with respect to h which is less than three are easily seen to contain an expression of the form $I_{sir} I_j I_p$, and their mathematical expectation is zero (see the third identity in (8.12)). For $s = 4, 5$, all terms containing I_{sir} have order at least three. These considerations imply that the inequality (8.15) holds. □

We now form the random vector $\overline{X}$ such that the inequalities

$$\left| \mathbf{E} \left(\prod_{j=1}^{s} \widetilde{\Delta}^{i_j} - \prod_{j=1}^{s} \overline{\Delta}^{i_j} \right) \right| \leq K(x) h^3, \qquad s = 1, \dots, 5, \qquad K(x) \in F, \tag{8.16}$$

as well as (8.14) hold. Then (8.15) and (8.16) imply (8.13). As a result we have constructed a vector $\overline{X}$ satisfying the inequalities (8.13) and (8.14).

We will construct $\overline{X}$ similar to $\widetilde{X}$ as follows:

$$\begin{aligned} \overline{X} = x &+ \sum_{r=1}^{q} \sigma_r \xi_r h^{1/2} + ah + \sum_{r=1}^{q} \sum_{i=1}^{q} \Lambda_i \sigma_r \xi_{ir} h \\ &+ \sum_{r=1}^{q} L\sigma_r \xi_r \cdot h^{3/2} + \sum_{r=1}^{q} (\Lambda_r a - L\sigma_r) \eta_r \cdot h^{3/2} + La \cdot \frac{h^2}{2}. \end{aligned} \tag{8.17}$$

LEMMA 8.4. *Under the conditions of Lemma 8.1, for the inequalities (8.13) and (8.14) to hold it suffices that the random variables ξ_r, ξ_{ir}, η_r in (8.17) have finite* *moments up to order six, inclusively, and that the following relations hold:*

$$\mathbf{E}\xi_r h^{1/2} = \mathbf{E}I_r = 0, \qquad \mathbf{E}\xi_{ir} h = \mathbf{E}I_{ir} = 0, \qquad \mathbf{E}\eta_r h^{3/2} = \mathbf{E}J_r = 0; \tag{8.18}$$

$$\mathbf{E}\xi_i\xi_r h = \mathbf{E}I_i I_r = \delta_{ir} h, \qquad \mathbf{E}\xi_i\xi_{rj} h^{3/2} = \mathbf{E}I_i I_{rj} = 0,$$

$$\mathbf{E}\xi_r\eta_j h^2 = \mathbf{E}I_r J_j = \delta_{rj}\frac{h^2}{2},$$

$$\mathbf{E}\xi_{ir}\xi_{js} h^2 = \mathbf{E}I_{ir} I_{js} = \begin{cases} \frac{h^2}{2} & \text{if } i = j,\ r = s, \\ 0 & \text{otherwise}, \end{cases}$$

$$\mathbf{E}\xi_{ir}\eta_j h^{5/2} = \mathbf{E}I_{ir} J_j = 0; \tag{8.19}$$

$$\mathbf{E}\xi_i\xi_r\xi_j h^{3/2} = \mathbf{E}I_i I_r I_j = 0,$$

$$\mathbf{E}\xi_i\xi_r\xi_{js} h^2 = \mathbf{E}I_i I_r I_{js} = \begin{cases} \frac{h^2}{2} & \text{if } j \neq s \text{ and either } i = j,\ r = s \text{ or } i = s,\ r = j, \\ h^2 & \text{if } i = r = j = s, \\ 0 & \text{otherwise}, \end{cases}$$

$$\mathbf{E}\xi_i\xi_r\eta_j h^{5/2} = \mathbf{E}I_i I_r I_j = 0, \qquad \mathbf{E}\xi_i\xi_{jr}\xi_{sl} h^{5/2} = \mathbf{E}I_i I_{jr} I_{sl} = 0; \tag{8.20}$$

$$\mathbf{E}\xi_i\xi_r\xi_j\xi_s h^2 = \mathbf{E}I_i I_r I_j I_s = \begin{cases} h^2 & \text{if } \{i,r,j,s\} \text{ consists of two pairs of equal numbers}, \\ 3h^2 & \text{if } i = r = j = s, \\ 0 & \text{otherwise}, \end{cases}$$

$$\mathbf{E}\xi_i\xi_r\xi_j\xi_{sl} h^{5/2} = \mathbf{E}I_i I_r I_j I_{sl} = 0; \tag{8.21}$$

$$\mathbf{E}\xi_i\xi_r\xi_j\xi_s\xi_l h^{5/2} = \mathbf{E}I_i I_r I_j I_s I_l = 0. \tag{8.22}$$

PROOF. Inequality (8.14) for the sixth moments of the moduli of the coordinates of the vector $\overline{\Delta} = \overline{X} - x$ clearly follows from (8.17), since each term in $\overline{\Delta}$ has at least order of smallness 1/2 with respect to h. Further, all identities (8.18)–(8.22) consist of two parts: a right part and a left part. We will prove the right parts below. The left parts of (8.18) clearly suffice for (8.16) to hold (and so also for (8.13)) for $s = 1$, i.e. such that up to $O(h^3)$ the first moments of the coordinates of the vectors Δ and $\overline{\Delta}$ coincide. The left parts of (8.18)–(8.19) suffice for the second moments; (8.18)–(8.20) suffice for the third moments; (8.18)–(8.21) suffice for the fourth moments; and (8.18)–(8.22) suffice for the fifth moments. Almost all right parts of (8.18)–(8.22) can be easily derived from oddness and independence considerations; only the computation of the mathematical expectations $\mathbf{E}I_{ir}I_{js}$ and $\mathbf{E}I_i I_r I_{js}$ presents some difficulties. Without loss of generality we set $t = 0$. To compute $\mathbf{E}I_{ir}I_{js} = \mathbf{E}\int_0^h w_i(\theta)\,dw_r(\theta)\cdot\int_0^h w_j(\theta)\,dw_s(\theta)$ we introduce the system of equations

$$dx(\theta) = w_i(\theta)\,dw_r(\theta), \qquad x(0) = 0,$$
$$dy(\theta) = w_j(\theta)\,dw_s(\theta), \qquad y(0) = 0.$$

It is obvious that $\mathbf{E}I_{ir}I_{js} = \mathbf{E}x(h)y(h)$. By Itô's formula we have

$$dxy = yw_i\,dw_r + xw_j\,dw_s + w_i w_j \delta_{rs}\,d\theta.$$

Therefore

$$d\mathbf{E}\left(x(\theta)y(\theta)\right) = \delta_{rs}\mathbf{E}\left(w_i(\theta)w_j(\theta)\right)\,d\theta,$$

which immediately implies the last of the identities (8.19).

To compute $\mathbf{E}I_iI_{js}I_r = \mathbf{E}w_i(h)w_r(h)\int_0^h w_j(\theta)\,dw_s(\theta)$ we introduce the equation

$$dy(\theta) = w_j(\theta)\,dw_s(\theta), \qquad y(0) = 0.$$

It is obvious that $\mathbf{E}I_iI_rI_{js} = \mathbf{E}w_i(h)w_r(h)y(h)$. By Itô's formula,

$$\begin{aligned} d(w_iw_ry) &= w_ry\,dw_i + w_iy\,dw_r + w_iw_rw_j\,dw_s \\ &\quad + y\delta_{ir}\,d\theta + w_rw_j\delta_{is}\,d\theta + w_iw_j\delta_{rs}\,d\theta. \end{aligned}$$

In view of $\mathbf{E}y(\theta) = 0$ we obtain

$$d\mathbf{E}(w_iw_ry) = \delta_{is}\mathbf{E}(w_rw_j)\,d\theta + \delta_{rs}\mathbf{E}(w_iw_j)\,d\theta,$$

which immediately implies the second identity in (8.20). □

8.3. Theorem on a method with one-step approximation of third order of accuracy.

THEOREM 8.1. *Suppose the conditions of Lemma* 8.1 *hold. Let $f(x)$ be a function belonging, with all its partial derivatives up to order six inclusively, to the class F. Let ξ_i, η_i, ξ_{ij} be chosen such that* (8.18)–(8.22) *hold. Then for $\overline{X}$ from* (8.17) *the following inequality holds (recall that $X = X(t+h)$):*

$$\left|\mathbf{E}f(X) - \mathbf{E}f(\overline{X})\right| \le K(x)h^3, \qquad K(x) \in F, \tag{8.23}$$

i.e. in the sense of weak approximations the method (8.17) *has third order of accuracy in a single step.*

PROOF. By Lemma 8.1–Lemma 8.4 the inequalities (8.13), (8.14) hold. Moreover, similarly to the proof of (8.14) in Lemma 8.4, we can prove the inequality

$$\mathbf{E}\prod_{j=1}^{s}\left|\Delta^{i_j}\right| \le K(x)h^3, \qquad i_j = 1,\dots,n, \qquad s = 6, \qquad K(x) \in F. \tag{8.24}$$

We now write down the Taylor expansion, in a neighborhood of x, of $f(X)$ with respect to powers of $\Delta^i = X^i - x^i$ and with Lagrange remainder term containing terms of order six. We proceed similarly for $f(\overline{X})$ with respect to the powers of $\overline{\Delta}^i = \overline{X}^i - x^i$. Using (8.13), (8.14), and (8.24) we arrive at (8.23). □

8.4. Modeling of random variables and constructive formation of a one-step approximation of third order of accuracy. There are various methods that satisfy the relations (8.18)–(8.22). We will look for random variables ξ_i, η_i, ξ_{ij} as follows. Consider mutually independent symmetric random variables ξ_i and ζ_j, $i, j = 1, \dots, q$, and put

$$\eta_i = \frac{1}{2}\xi_i, \qquad \xi_{ij} = \frac{1}{2}\xi_i\xi_j - \frac{1}{2}\gamma_{ij}\zeta_i\zeta_j, \qquad \gamma_{ij} = \begin{cases} -1, & i < j, \\ 1, & i \geq j. \end{cases} \tag{8.25}$$

We will assume that ξ_i and ζ_j have all moments needed. Below we will verify that if, in addition to the above-said, we require that $\mathbf{E}\xi_i^2 = 1$, $\mathbf{E}\xi_i^4 = 3$, $\mathbf{E}\zeta_i^2 = 1$, $\mathbf{E}\zeta_i^4 = 1$, then all the relations (8.18)–(8.22) are satisfied. For example, we can model the ξ_i by the law $N(0,1)$ and the ζ_i by the law $\mathbf{P}(\zeta = -1) = \mathbf{P}(\zeta = 1) = 1/2$. But ξ_i can be modeled by a much simpler law too. For example, by the law $\mathbf{P}(\xi = 0) = 2/3$, $\mathbf{P}(\xi = -\sqrt{3}) = \mathbf{P}(\xi = \sqrt{3}) = 1/6$. We turn to the immediate verification of (8.18)–(8.22) for the random variables (8.25). We repeat that ξ_i and ζ_j are independent variables for which

$$\begin{aligned} \mathbf{E}\xi_i = \mathbf{E}\xi_i^3 = \mathbf{E}\xi_i^5 = 0, \qquad & \mathbf{E}\xi_i^2 = 1, \qquad \mathbf{E}\xi_i^4 = 3, \\ \mathbf{E}\zeta_i = \mathbf{E}\zeta_i^3 = 0, \qquad & \mathbf{E}\zeta_i^2 = \mathbf{E}\zeta_i^4 = 1. \end{aligned} \tag{8.26}$$

Vanishing of the odd moments is a consequence of the above-mentioned symmetry. Indeed, only independence and (8.26) are required, and the requirement of symmetry is simply a convenient sufficient condition for the vanishing of the odd moments.

Many of the relations (8.18)–(8.22) can be verified rather simply. Therefore we only treat the verification of the more complicated ones.

LEMMA 8.5. *Let ξ_i, ζ_j be independent random variables such that* (8.25), (8.26) *hold. Then*

$$\mathbf{E}\xi_{ir}\xi_{js} = \begin{cases} \frac{1}{2} & \textit{if } i = j,\ r = s, \\ 0 & \textit{otherwise}. \end{cases} \tag{8.27}$$

PROOF. We have

$$\mathbf{E}\xi_{ir}\xi_{js} = \frac{1}{4}\left(\mathbf{E}\xi_i\xi_r\xi_j\xi_s - \gamma_{js}\mathbf{E}\xi_i\xi_r\zeta_j\zeta_s - \gamma_{ir}\mathbf{E}\xi_j\xi_s\zeta_i\zeta_r + \gamma_{ir}\gamma_{js}\mathbf{E}\zeta_i\zeta_r\zeta_j\zeta_s\right). \tag{8.28}$$

Let $i \neq j$. If r is not equal to i and not equal to j, then, clearly, all mathematical expectations at the righthand side of (8.8) are zero. So, for $i \neq j$ the righthand side can be nonzero only if $r = i$ or $r = j$. Consider the case $i \neq j$, $r = i$. If also $s \neq j$, then the righthand side is zero. Therefore we may assume that in (8.28) $i \neq j$, $r = i$, $s = j$. Then $\gamma_{ir} = \gamma_{js} = 1$, and each of the four mathematical expectations at the righthand side of (8.28) is equal to 1. As a result, for $i \neq j$, $r = i$ the righthand side vanishes. Consider now the case $i \neq j$, $r = j$. In this case the righthand side of (8.28) can be nonzero only if $s = i$. So, let $i \neq j$, $r = j$, $s = i$. We have $\mathbf{E}\xi_i\xi_r\xi_j\xi_s = \mathbf{E}\zeta_i\zeta_r\zeta_j\zeta_s = 1$, but $\mathbf{E}\xi_i\xi_r\zeta_j\zeta_s = \mathbf{E}\xi_j\xi_s\zeta_i\zeta_r = 0$. For $i \neq j$, $r = j$, $s = i$ the product $\gamma_{ir}\gamma_{js}$ is always -1. In fact, if $i < j$ then $i < r$ and $j > s$, since by assumption $r = j$, $s = i$. But

$\gamma_{ir} = -1$ for $i < r$ and $\gamma_{js} = 1$ for $j > s$. Hence $\gamma_{ir}\gamma_{js} = -1$. The case $i > j$ can be similarly treated. So, if $i \neq j$ the righthand side of (8.28) is always zero.

Let now $i = j$. Then the righthand side of (8.28) can only be nonzero for $r = s$. We distinguish three cases. In the first case $i < r$. Then $j < s$, $\gamma_{ir} = \gamma_{js} = -1$, $\mathbf{E}\xi_i\xi_r\xi_j\xi_s = \mathbf{E}\zeta_i\zeta_r\zeta_j\zeta_s = 1$, $\mathbf{E}\xi_i\xi_r\zeta_j\zeta_s = \mathbf{E}\xi_j\xi_s\zeta_i\zeta_r = 0$, and hence $\mathbf{E}\xi_{ir}\xi_{js} = 1/2$. The second case, $i > r$, can be similarly treated. It differs by the relations $\gamma_{ir} = \gamma_{js} = 1$. The third case, $i = r$, gives $i = j = r = s$, $\mathbf{E}\xi_i\xi_r\xi_j\xi_s = 3$, $\mathbf{E}\xi_i\xi_r\zeta_j\zeta_s = \mathbf{E}\xi_j\xi_s\zeta_i\zeta_r = \mathbf{E}\zeta_i\zeta_r\zeta_j\zeta_s = 1$, $\gamma_{ir} = \gamma_{js} = 1$, and hence $\mathbf{E}\xi_{ir}\xi_{js} = 1/2$. □

We need this Lemma to substantiate the forelast relation in (8.19). The other relations in (8.19), as well as (8.18), can be verified in an obvious manner. In (8.20) the second relation presents some difficulty. To verify it we prove the following Lemma.

LEMMA 8.6. *Let ξ_i, ζ_j be independent random variables and suppose* (8.25)–(8.26) *hold. Then*

$$\mathbf{E}\xi_i\xi_r\xi_{js} = \begin{cases} \frac{1}{2} & \text{if } j \neq s \text{ and either } i = j,\ r = s \text{ or } i = s,\ r = j, \\ 1 & \text{if } i = r = j = s, \\ 0 & \text{otherwise.} \end{cases} \tag{8.29}$$

PROOF. We have

$$\mathbf{E}\xi_i\xi_r\xi_{js} = \frac{1}{2}\left(\mathbf{E}\xi_i\xi_r\xi_j\xi_s - \gamma_{js}\mathbf{E}\xi_i\xi_r\zeta_j\zeta_s\right). \tag{8.30}$$

For $j \neq s$ the righthand side of (8.30) can be nonzero only for $i = j$, $r = s$ or $i = s$, $r = j$. In both these cases $\mathbf{E}\xi_i\xi_r\xi_j\xi_s = 1$, $\mathbf{E}\xi_i\xi_r\zeta_j\zeta_s = 0$, which proves (8.29) for $j \neq s$. If $j = s$ but $i \neq j$, then the righthand side of (8.30) can be nonzero only if $i = r$. But in this case $\gamma_{js} = 1$, $\mathbf{E}\xi_i\xi_r\xi_j\xi_s = \mathbf{E}\xi_i\xi_r\zeta_j\zeta_s = 1$ and, hence, the righthand side of (8.30) is zero. Let $j = s$, $i = j$. Then (8.30) can be nonzero only if $i = r$, i.e. $i = r = j = s$. If $i = r = j = s$, then $\mathbf{E}\xi_i\xi_r\xi_j\xi_s = 3$, $\mathbf{E}\xi_i\xi_r\zeta_j\zeta_s = 1$, $\gamma_{js} = 1$, i.e. $\mathbf{E}\xi_i\xi_r\xi_{js} = 1$. □

The other relations (8.20)–(8.22) can be verified in a simple way. As a result we can write the *one-step approximation* (8.17) as

$$\begin{aligned}\overline{X} = x &+ \sum_{r=1}^{q}\sigma_r\xi_r h^{1/2} + ah + \sum_{r=1}^{q}\sum_{i=1}^{q}\Lambda_i\sigma_i\xi_{ir}h \\ &+ \frac{1}{2}\sum_{r=1}^{q}\left(\Lambda_r a + L\sigma_r\right)\xi_r h^{3/2} + La\cdot\frac{h^2}{2},\end{aligned} \tag{8.31}$$

where ξ_{ir} satisfies (8.25), and ξ_i, ζ_j are independent random variables satisfying (8.26). In particular, ξ_i can be modeled by the law $\mathbf{P}(\xi = 0) = 2/3$, $\mathbf{P}(\xi = \sqrt{3}) = \mathbf{P}(\xi = -\sqrt{3}) = 1/6$, and ζ_j can be modeled by $\mathbf{P}(\zeta = -1) = \mathbf{P}(\zeta = 1) = 1/2$. The one-step approximation (8.31) has third order of accuracy in the sense of weak approximation.

9. The main theorem on convergence of weak approximations and methods of order of accuracy two

9.1. A theorem on the relation between one-step approximation and approximation on a finite interval. Next to the system (8.1) we consider the approximation

$$\overline{X}_{t,x}(t+h) = x + A(t,x,h;\xi), \tag{9.1}$$

where ξ is a random variable (in general, a vector) having moments of sufficiently high order, and A is a vector function of dimension n. Partition the interval $[t_0, t_0+T]$ into N equal parts, with step $h = T/N$: $t_0 < t_1 < \cdots < t_N = t_0 + T$, $t_{k+1} - t_k = h$. From (9.1) we construct the sequence

$$\overline{X}_0 = X_0 = X(t_0), \qquad \overline{X}_{k+1} = \overline{X}_k + A(t_k, \overline{X}_k, h; \xi_k), \quad k = 0, \dots, N-1, \tag{9.2}$$

where ξ_0 is independent of $\overline{X}_0$, while ξ_k for $k > 0$ is independent of $\overline{X}_0, \dots, \overline{X}_k$, $\xi_0, \dots, \xi_{k-1}$. As before, we write $\Delta = X - x = X_{t,x}(t+h) - x$, $\overline{\Delta} = \overline{X} - x = \overline{X}_{t,x}(t+h) - x$, let $X(t) = X_{t_0,X_0}(t)$ be a solution of (9.1), and $\overline{X}_{t_0,X_0}(t_k) = \overline{X}_k$.

THEOREM 9.1. *Suppose that the following conditions hold:*

1) *the coefficients of equation* (8.1) *are continuous, satisfy a Lipschitz condition* (8.2) *and, together with their partial derivatives with respect to x and of order up to $2p+2$, inclusively, belong to F;*
2) *the method* (9.1) *is such that*

$$\left| \mathbf{E} \left(\prod_{j=1}^{s} \Delta^{i_j} - \prod_{j=1}^{s} \overline{\Delta}^{i_j} \right) \right| \le K(x) h^{p+1}, \qquad s = 1, \dots, 2p+1, \qquad K(x) \in F, \tag{9.3}$$

$$\mathbf{E} \prod_{j=1}^{2p+2} \left| \overline{\Delta}^{i_j} \right| < K(x) h^{p+1}, \qquad K(x) \in F; \tag{9.4}$$

3) *the function $f(x)$, together with its partial derivatives with respect to x and of order up to $2p_2+2$, inclusively, belong to F;*
4) *for sufficiently large m (specified below) the $\mathbf{E}\left|\overline{X}_k\right|^{2m}$ exist and are uniformly bounded with respect to N and $k = 0, 1, \dots, N$.*

Then, for all N and all $k = 0, 1, \dots, N$ the following inequality holds:

$$\left| \mathbf{E} f(X_{t_0,X_0}(t_k)) - \mathbf{E} f(\overline{X}_{t_0,X_0}(t_k)) \right| \le K h^p, \tag{9.5}$$

i.e. the method (9.2) *has order of accuracy p in the sense of weak approximations.*

PROOF. We first of all note that the Lipschitz condition (8.2) implies that for any $m > 0$ the mathematical expectations $\mathbf{E}\left|X(\theta)\right|^{2m}$ exist and are uniformly bounded with respect to $\theta \in [t_0, t_0+T]$, if only $\mathbf{E}\left|X(t_0)\right|^{2m} < \infty$ (see [8, p. 48]). Moreover, the same (8.2) implies

$$\mathbf{E} \prod_{j=1}^{2p+2} \left| \Delta^{i_j} \right| < K(x) h^{p+1}, \qquad K(x) \in F. \tag{9.6}$$

Further, suppose that $u(x)$ is a function that, together with its partial derivatives with respect to x and of order up to $2p+2$, inclusively, belong to F. Then

$$\left|\mathbf{E}u\left(X_{t,x}(t+h)\right)-\mathbf{E}u\left(\overline{X}_{t,x}(t+h)\right)\right|\leq K(x)h^{p+1},\qquad K(x)\in F. \tag{9.7}$$

Thanks to (9.3), (9.4), (9.6), the proof of (9.7) is completely similar to the proof of Theorem 8.1. We introduce the function

$$u(s,x)=\mathbf{E}f(X_{s,x}(t_{k+1})).$$

By requirements 1) and 3), u has partial derivatives with respect to x of order up to $2p+2$, inclusively; moreover, as for u, these derivatives belong to F (see [8, pp. 60–61]). Therefore the function $u(s,x)$ satisfies an estimate of the form (9.7) uniformly with respect to $s\in[t_0,t_{k+1}]$.

Further, since $\overline{X}_0=X_0$, $X_{t_0,X_0}(t_1)=X(t_1)$, $X_{t_1,X_{t_0,\overline{X}_0}(t_1)}(t_{k+1})=X(t_{k+1})$, we have

$$\begin{aligned}\mathbf{E}f\left(X(t_{k+1})\right)=\mathbf{E}f\left(X_{t_1,X_{t_0,\overline{X}_0}(t_1)}(t_{k+1})\right)-\mathbf{E}f\left(X_{t_1,\overline{X}_1}(t_{k+1})\right)\\+\mathbf{E}f\left(X_{t_1,\overline{X}_1}(t_{k+1})\right).\end{aligned} \tag{9.8}$$

Similarly, since $X_{t_1,\overline{X}_1}(t_{k+1})=X_{t_2,X_{t_1,\overline{X}_1}(t_2)}(t_{k+1})$, we have

$$\begin{aligned}\mathbf{E}f\left(X_{t_1,\overline{X}_1}(t_{k+1})\right)=\mathbf{E}f\left(X_{t_2,X_{t_1,\overline{X}_1}(t_2)}(t_{k+1})\right)-\mathbf{E}f\left(X_{t_2,\overline{X}_2}(t_{k+1})\right)\\+\mathbf{E}f\left(X_{t_2,\overline{X}_2}(t_{k+1})\right).\end{aligned} \tag{9.9}$$

Now (9.8) and (9.9) imply

$$\begin{aligned}\mathbf{E}f\left(X(t_{k+1})\right)&=\mathbf{E}f\left(X_{t_1,X_{t_0,\overline{X}_0}(t_1)}(t_{k+1})\right)-\mathbf{E}f\left(X_{t_1,\overline{X}_1}(t_{k+1})\right)\\&+\mathbf{E}f\left(X_{t_2,X_{t_1,\overline{X}_1}(t_2)}(t_{k+1})\right)-\mathbf{E}f\left(X_{t_2,\overline{X}_2}(t_{k+1})\right)\\&+\mathbf{E}f\left(X_{t_2,\overline{X}_2}(t_{k+1})\right).\end{aligned}$$

Proceeding further, we obtain

$$\begin{aligned}\mathbf{E}f\left(X(t_{k+1})\right)=\sum_{i=0}^{k-1}\left(\mathbf{E}f\left(X_{t_{i+1},X_{t_i,\overline{X}_i}(t_{i+1})}(t_{k+1})\right)-\mathbf{E}f\left(X_{t_{i+1},\overline{X}_{i+1}}(t_{k+1})\right)\right)\\+\mathbf{E}f\left(X_{t_k,\overline{X}_k}(t_{k+1})\right).\end{aligned} \tag{9.10}$$

This immediately implies the identity (recall that $\overline{X}_{i+1}=\overline{X}_{t_i,\overline{X}_i}(t_{i+1})$)

$$\begin{aligned}\mathbf{E}f\left(X(t_{k+1})\right)-\mathbf{E}f\left(\overline{X}_{k+1}\right)=\sum_{i=0}^{k-1}\left(\mathbf{E}\mathbf{E}\left(f\left(X_{t_{i+1},X_{t_i,\overline{X}_i}(t_{i+1})}(t_{k+1})\right)\mid X_{t_i,\overline{X}_i}(t_{i+1}\right)\right)\\-\mathbf{E}\mathbf{E}\left(f\left(X_{t_{i+1},\overline{X}_{t_i,\overline{X}_i}(t_{i+1})}(t_{k+1})\right)\mid X_{t_i,\overline{X}_i}(t_{i+1})\right)\\+\mathbf{E}f\left(X_{t_k,\overline{X}_k}(t_{k+1})\right)-\mathbf{E}f\left(\overline{X}_{t_k,\overline{X}_k}(t_{k+1})\right).\end{aligned} \tag{9.11}$$

According to the definition of $u(s,x)$, (9.11) implies

$$\begin{aligned}
&\left|\mathbf{E}f\left(X(t_{k+1})\right)-\mathbf{E}f\left(\overline{X}_{k+1}\right)\right| \\
&\quad=\left|\sum_{i=0}^{k-1}\left(\mathbf{E}u\left(t_{i+1},X_{t_i,\overline{X}_i}(t_i+h)\right)-\mathbf{E}u\left(t_{i+1},\overline{X}_{t_i,\overline{X}_i}(t_i+h)\right)\right)\right. \\
&\qquad\left.+\left(\mathbf{E}f\left(X_{t_k,\overline{X}_k}(t_{k+1})\right)-\mathbf{E}f\left(\overline{X}_{t_k,\overline{X}_k}(t_{k+1})\right)\right)\right| \\
&\quad\le\sum_{i=0}^{k-1}\mathbf{E}\left|\mathbf{E}\left(u\left(t_{i+1},X_{t_i,\overline{X}_i}(t_i+h)\right)-u\left(t_{i+1},\overline{X}_{t_i,\overline{X}_i}(t_i+h)\right)\mid\overline{X}_i\right)\right| \\
&\qquad+\mathbf{E}\left|\mathbf{E}\left(f\left(X_{t_k,\overline{X}_k}(t_{k+1})\right)-\mathbf{E}f\left(\overline{X}_{t_k,\overline{X}_k}(t_{k+1})\right)\mid\overline{X}_k\right)\right|.
\end{aligned}\tag{9.12}$$

We now note that the functions $u(s,x)$ and $f(x)$, which belong to F and so satisfy an inequality of the form (9.7), satisfy also the conditional version of this inequality. Suppose that for both $u(s,x)$ and $f(x)$ we have a function $K(x)$ in this inequality with $\kappa=2m$. Then (9.12) implies

$$\left|\mathbf{E}f\left(X(t_{k+1})\right)-\mathbf{E}f\left(\overline{X}_{k+1}\right)\right|\le\sum_{i=0}^{k-1}K\left(1+\mathbf{E}\left|\overline{X}_i\right|^{2m}\right)h^{p+1}+K\left(1+\mathbf{E}\left|\overline{X}_k\right|^{2m}\right)h^{p+1}.$$

Assuming that requirement 4) holds for precisely this $2m$, we arrive at (9.5). □

We will now give a sufficient condition for requirement 4) in Theorem 9.1 which is convenient in practice.

LEMMA 9.1. *Suppose that for $h<1$,*

$$|\mathbf{E}A(t_k,x,h;\xi_k)|\le K\left(1+|x|\right)h,\tag{9.13}$$

$$|A(t_k,x,h;\xi_k)|\le M(\xi_k)\left(1+|x|\right)h^{1/2},\tag{9.14}$$

where $M(\xi_k)$ has moments of all orders.

Then for every even number $2m$ the mathematical expectations $\mathbf{E}\left|\overline{X}_k\right|^{2m}$ exist and are uniformly bounded with respect to N and $k=1,\dots,N$, if only $\mathbf{E}\left|\overline{X}_0\right|^{2m}$ exists.

PROOF. For the ith coordinate of the vector $\overline{X}_{k+1}$ we have

$$\begin{aligned}
\left(\overline{X}_{k+1}^i\right)^{2m}&=\left(\overline{X}_k^i+A^i(t_k,\overline{X}_k,h;\xi_k)\right)^{2m} \\
&=\left(\overline{X}_k^i\right)^{2m}+C_{2m}^1\left(\overline{X}_k^i\right)^{2m-1}A^i(t_k,\overline{X}_k,h;\xi_k) \\
&\quad+\sum_{j=2}^{2m}C_{2m}^j\left(\overline{X}_k^i\right)^{2m-j}\left(A^i(t_k,\overline{X}_k,h;\xi_k)\right)^j.
\end{aligned}\tag{9.15}$$

Using (9.13), we obtain

$$\begin{aligned}\left|\mathbf{E}\left(\overline{X}_k^i\right)^{2m-1} A^i(t_k,\overline{X}_k,h;\xi_k)\right| &= \left|\mathbf{E}\left(\left(\overline{X}_k^i\right)^{2m-1}\mathbf{E}\left(A^i(t_k,\overline{X}_k,h;\xi_k)\right)\overline{X}_k\right)\right| \\ &\le \left|\mathbf{E}\left(\overline{X}_k^i\right)^{2m-1} K\left(1+\left|\overline{X}_k\right|\right)h\right| \\ &\le K\left(1+\mathbf{E}\left|\overline{X}_k\right|^{2m}\right)h. \end{aligned} \tag{9.16}$$

By (9.14), for $h<1$ and $j=2,\dots,2m$ we obtain

$$\begin{aligned}\left|\mathbf{E}\left|\overline{X}_k^i\right|^{2m-j}\left(A^i(t_k,\overline{X}_k,h;\xi_k)\right)^j\right| &\le \mathbf{E}\left(\left|\overline{X}_k^i\right|^{2m-j}(M(\xi_k))^j\left(1+\left|\overline{X}_k\right|\right)^j h^{j/2}\right) \\ &\le K\left(1+\mathbf{E}\left|\overline{X}_k\right|^{2m}\right)h. \end{aligned} \tag{9.17}$$

Because of (9.15)–(9.17) and the inequality $|x|^{2m} \le K\sum_{i=1}^m (x^i)^{2m}$, where the constant K depends on n and m only, we obtain

$$\mathbf{E}\sum_{i=1}^n\left(\overline{X}_{k+1}^i\right)^{2m} \le \mathbf{E}\sum_{i=1}^n\left(\overline{X}_k^i\right)^{2m} + K\left(1+\sum_{i=1}^n\left(\overline{X}_k^i\right)^{2m}\right)\cdot h.$$

Using Lemma 1.3, this concludes the proof of Lemma 9.1. □

9.2. Theorem on a method of order of accuracy two. Theorem 9.1 and Lemma 9.1 imply a theorem on the order of accuracy of the method

$$\begin{aligned} X_{k+1} = X_k &+ \sum_{r=1}^q \sigma_{rk}\xi_{rk}h^{1/2} + a_k h + \sum_{r=1}^q\sum_{i=1}^q (\Lambda_r\sigma_r)_k\,\xi_{irk}h \\ &+ \frac{1}{2}\sum_{r=1}^q (L\sigma_r+\Lambda_r a)_k\,\xi_{rk}h^{3/2} + (La)_k\frac{h^2}{2}, \end{aligned} \tag{9.18}$$

which is constructed according to (8.31).

In (9.18) the coefficients σ_{rk}, a_k, $(\Lambda_i\sigma_r)_k$, etc. are computed at the point (t_k, X_k), and the sets of random variables ξ_{rk}, ξ_{irk} are independent for distinct k and can, for each k, be modeled as in (8.31).

THEOREM 9.2. *Suppose the conditions of Lemma* 8.1 *hold. Suppose also that the functions $\Lambda_i\sigma_r$, $L\sigma_r$, $\Lambda_r a$, and La grow at most as a linear function in $|x|$ as $|x|$ grows (the functions a and σ_r satisfy this requirement thanks to the Lipschitz condition* (8.2)), *i.e.* (9.13)–(9.14) *hold for* (9.18). *Then the method* (9.18) *has order of accuracy two in the sense of weak approximation, i.e. for a sufficiently large class of functions f we have* (9.5) *with $p = 2$ (under the conditions of this Theorem, this class of functions contains the functions that belong, together with their partial derivatives up to order six, inclusively, to F).*

The proof of this theorem clearly follows from the properties of the one-step approximation (8.31) proved in §8, Lemma 9.1, and Theorem 9.1.

EXAMPLE 9.1. Consider the one-dimensional equation (8.1) with a single noise, i.e. $q = 1$. In this case

$$\xi_{11} = \frac{1}{2}\left(\xi^2 - 1\right),$$

where ξ is, e.g., $N(0,1)$-distributed or distributed by the law $\mathbf{P}(\xi = 0) = 2/3$, $\mathbf{P}(\xi = -\sqrt{3}) = \mathbf{P}(\xi = \sqrt{3}) = 1/6$. Formula (9.18) takes the form

$$\begin{aligned} X_{k+1} = X_k + \sigma_k \xi_k h^{1/2} + a_k h + \frac{1}{2}\left(\sigma \frac{\partial \sigma}{\partial x}\right)_k \left(\xi_k^2 - 1\right) h \\ + \frac{1}{2}\left(\frac{\partial \sigma}{\partial t} + a\frac{\partial \sigma}{\partial x} + \frac{1}{2}\sigma^2 \frac{\partial^2 \sigma}{\partial x^2} + \sigma \frac{\partial a}{\partial x}\right) \xi_k h^{3/2} \\ + \left(\frac{\partial a}{\partial t} + a\frac{\partial a}{\partial x} + \frac{1}{2}\sigma^2 \frac{\partial^2 a}{\partial x^2}\right)_k \frac{h^2}{2}. \end{aligned} \tag{9.19}$$

This formula has been derived in [26] for the first time ever, using Taylor expansions of the characteristic functions of the variables $\Delta = X_{t,x}(t+h) - x$ and $\overline{\Delta} = \overline{X}_{t,x}(t+h) - x$.

9.3. Runge–Kutta type methods. The method (9.19) may present considerable difficulties because of the necessity of computing, at each step, the derivatives of the coefficients a and σ. Using the idea of the *Runge–Kutta method*, one can propose a number of ways in which, by recomputation, one can obtain a method not including all the derivatives participating in (9.19). We give one concrete, sufficiently simple method of this kind (it was proposed in [26]):

$$\begin{aligned} X_{k+1} = X_k + \frac{1}{2}\sigma_k \xi_k h^{1/2} + \frac{1}{2}\left(a - \sigma \frac{\partial \sigma}{\partial x}\right)_k h + \frac{1}{2}\left(\sigma \frac{\partial \sigma}{\partial x}\right)_k \xi_k^2 h \\ + \frac{1}{2} a(t_k + h, X_k + a_k h + \sigma_k \xi_k h^{1/2}) h \\ + \frac{1}{4}\left(\sigma\left(t_k + h, X_k + a_k h + \sigma_k \xi_k \left(\frac{h}{3}\right)^{1/2}\right)\right. \\ \left. + \sigma\left(t_k + h, X_k + a_k h - \sigma_k \xi_k \left(\frac{h}{3}\right)^{1/2}\right)\right) \xi_k h^{1/2}, \end{aligned} \tag{9.20}$$

where ξ_k are the same variables as in (9.19).

To get convinced of the fact that the method (9.20) is a method of order of accuracy two, we note that

$$\frac{1}{2}a(t+h, x+ah+\sigma\xi h^{1/2})h$$

$$= \frac{1}{2}\left(a + \frac{\partial a}{\partial t}h + \frac{\partial a}{\partial x}ah + \frac{\partial a}{\partial x}\sigma\xi h^{1/2} + \frac{1}{2}\frac{\partial^2 a}{\partial x^2}\sigma^2\xi^2 h\right)h \tag{9.21}$$

$$+ \frac{1}{2}\frac{\partial^2 a}{\partial x^2}a\sigma\xi h^{5/2} + \frac{1}{2}\frac{\partial^2 a}{\partial t\partial x}\sigma\xi h^{5/2} + O(h^3),$$

$$\frac{1}{4}\left(\sigma\left(t+h, x+ah+\sigma\xi\left(\frac{h}{3}\right)^{1/2}\right) + \sigma\left(t+h, x+ah-\sigma\xi\left(\frac{h}{3}\right)^{1/2}\right)\right)\xi h^{1/2}$$

$$= \frac{1}{4}\left(2\sigma + 2\frac{\partial\sigma}{\partial t}h + 2\frac{\partial\sigma}{\partial x}ah + \frac{1}{3}\frac{\partial^2\sigma}{\partial x^2}\sigma^2\xi^2 h\right)\xi h^{1/2}$$

$$+ \frac{1}{2}\frac{\partial^2\sigma}{\partial x\partial t}a\xi h^{5/2} + \frac{1}{12}\frac{\partial^3\sigma}{\partial t\partial x^2}\sigma^2\xi^3 h^{5/2} + \frac{1}{4}\frac{\partial^2\sigma}{\partial x^2}a^2\xi h^{5/2} \tag{9.22}$$

$$+ \frac{1}{12}\frac{\partial^3\sigma}{\partial x^3}a\sigma^2\xi^3 h^{5/2} + \frac{1}{432}\frac{\partial^4\sigma}{\partial x^4}\sigma^4\xi^5 h^{5/2} + O(h^3).$$

Substituting (9.22) and (9.21) into (9.20) we see that X_{k+1} in (9.20) differs from X_{k+1} in (9.21), first by the sum

$$s_1 = \left(\frac{1}{2}\frac{\partial^2 a}{\partial x^2}a\sigma\xi_k + \frac{1}{2}\frac{\partial^2 a}{\partial t\partial x}\sigma\xi_k + \frac{1}{2}\frac{\partial^2\sigma}{\partial t\partial x}a\xi + \frac{1}{12}\frac{\partial^3\sigma}{\partial t\partial x^2}\sigma^2\xi_k^3\right.$$
$$\left. + \frac{1}{4}\frac{\partial^2\sigma}{\partial x^2}a^2\xi_k + \frac{1}{12}\frac{\partial^3\sigma}{\partial x^3}a\sigma^2\xi_k^3 + \frac{1}{432}\frac{\partial^4\sigma}{\partial x^4}\sigma^4\xi_k^5\right)h^{5/2} + O(h^3),$$

secondly, the term

$$\frac{1}{4}\sigma^2\frac{\partial^2\sigma}{\partial x^2}\xi_k h^{3/2}$$

in (9.19) must be replaced by

$$\frac{1}{12}\sigma^2\frac{\partial^2\sigma}{\partial x^2}\xi_k^3 h^{3/2},$$

and, thirdly, the term

$$\frac{1}{4}\sigma^2\frac{\partial^2 a}{\partial x^2}h^2$$

in (9.19) must be replaced by

$$\frac{1}{4}\sigma^2\frac{\partial^2 a}{\partial x^2}\xi_k^2 h^2.$$

It is easy to see that these differences have no influence on the fulfillment of the conditions of the main Theorem 9.1 with $p = 2$.

Thus, the method (9.20) has order of accuracy two. At each step it requires two recomputations of the function a, three recomputations of the function σ, one recomputation of the function $\partial\sigma/\partial x$, and the modeling of the single random variable ξ.

In [59] another *Runge–Kutta type method* has been proposed for autonomous equations (see also [54]). We give the scheme of this method. Here one first constructs $X_{k+1/2}$ by the formula

$$X_{k+1/2} = X_k + \sigma_k \xi_k^{(1)} \sqrt{\frac{h}{2}} + \left(a - \frac{1}{2}\sigma\sigma'\right)_k \frac{h}{2} + \frac{1}{2} (\sigma\sigma')_k \xi_k^{(1)2} \frac{h}{2}, \tag{9.23}$$

and subsequently obtains X_{k+1}:

$$\begin{aligned} X_{k+1} = X_k &+ 2\sigma_k \xi_k^{(1)} \sqrt{\frac{h}{2}} + 2\sigma_{k+1/2} \xi_k^{(2)} \sqrt{\frac{h}{2}} - \sigma_k \eta_k \sqrt{h} \\ &+ \left(a - \frac{1}{2}\sigma\sigma'\right)_{k+1/2} h + (\sigma\sigma')_k \xi_k^{(1)2} \frac{h}{2} \\ &+ (\sigma\sigma')_{k+1/2} \xi_{k+1}^{(2)2} \frac{h}{2} - \frac{1}{2} (\sigma\sigma')_k \eta_k^2 h. \end{aligned} \tag{9.24}$$

In (9.23) and (9.24), the random variables $\xi_k^{(1)}$ and $\xi_k^{(2)}$ are independent in total (with respect to the superscripts $^{(1)}$ and $^{(2)}$, and with respect to k), and each of them is distributed by a law with moments: $\mathbf{E}\xi = \mathbf{E}\xi^3 = \mathbf{E}\xi^5 = 0$, $\mathbf{E}\xi^2 = 1$, $\mathbf{E}\xi^4 = 3$, $\mathbf{E}\xi^6 < \infty$; the random variable η_k is defined as $\eta_k = \xi_k^{(1)} + \xi_k^{(2)}$.

At each step, the method (9.23)–(9.24) requires two recomputations of a, two recomputations of σ, two recomputations of σ', and the modeling of two random variables. In [54], [59] a generalisation of the method (9.23)–(9.24) to systems of stochastic differential equations is given. Note that both the method (9.20) and the method (9.23)–(9.24) require the computation of $\partial\sigma/\partial x$ at each step. The problem of constructing a sufficiently efficient method in which no derivatives would have to be computed needs its own, separate solution.

10. A method of order of accuracy three for systems with additive noises

Consider the *system of stochastic differential equations with additive noises*

$$dX = a(t, X)\, dt + \sum_{r=1}^{q} \sigma_r(t)\, dw_r(t). \tag{10.1}$$

Since the σ_r do not depend on x, numerical integration methods are essentially simpler for such systems. So, since in this case the $\Lambda_i\sigma_r$ vanish, the terms ξ_{irk} in the method (9.18) (which has order of accuracy two) are absent, and consequently we only have to model, at each step, the random variables ξ_{rk}. The method (9.18) takes the following form for the system (10.1):

$$\begin{aligned} X_{k=1} = X_k &+ \sum_{r=1}^{q} \sigma_r(t_k) \xi_{rk} h^{1/2} + a_k h \\ &+ \frac{1}{2} \sum_{r=1}^{q} \left(\sigma_r' + \left(\sigma_r, \frac{\partial}{\partial x}\right) a\right)_k \xi_{rk} h^{3/2} + (La)_k \frac{h^2}{2}. \end{aligned} \tag{10.2}$$

While for systems of general form the attempt to arrive at a method of order of accuracy three meets with extremely awkward constructions, for systems with additive noises the problem of constructing such a method can be solved relatively simply.

10.1. Main lemmas. To construct a method of order of accuracy three we write down the following formula for the solution $X_{t,x}(\theta) = X(\theta)$ of (10.1):

$$\begin{aligned}
X(t+h) &= x + \sum_{r=1}^{q} \sigma_r \int_t^{t+h} dw_r(\theta) + ah \\
&+ \sum_{r=1}^{q} \Lambda_r a \int_t^{t+h} (w_r(\theta) - w_r(t))\, d\theta + \sum_{r=1}^{q} \sigma_r' \int_t^{t+h} (\theta - t)\, dw_r(\theta) \\
&+ La \cdot \frac{h^2}{2} + \sum_{r=1}^{q}\sum_{i=1}^{q} \Lambda_i \Lambda_r a \int_t^{t+h} \left(\int_t^{\theta} (w_i(\theta_1) - w_i(t))\, dw_r(\theta_1) \right) d\theta \\
&+ \sum_{r=1}^{q}\sum_{i=1}^{q}\sum_{s=1}^{q} \Lambda_s \Lambda_i \Lambda_r a \int_t^{t+h} \left(\int_t^{\theta} \left(\int_t^{\theta_1} (w_s(\theta_2) - w_s(t))\, dw_i(\theta_2) \right) dw_r(\theta_1) \right) d\theta \\
&+ \sum_{r=1}^{q} \sigma_r'' \int_t^{t+h} \left(\int_t^{\theta} (\theta_1 - t)\, d\theta_1 \right) dw_r(\theta) \\
&+ \sum_{r=1}^{q} L\Lambda_r a \int_t^{t+h} \left(\int_t^{\theta} (\theta_1 - t)\, dw_r(\theta_1) \right) d\theta \\
&+ \sum_{r=1}^{q} \Lambda_r La \int_t^{t+h} \left(\int_t^{\theta} (w_r(\theta_1) - w_r(t))\, d\theta_1 \right) d\theta + L^2 a \cdot \frac{h^3}{6} + \rho. \qquad (10.3)
\end{aligned}$$

In (10.3) all coefficients σ_r, $\Lambda_r a$, σ_r', La, $\Lambda_i\Lambda_r a$, $\Lambda_s\Lambda_i\Lambda_r a$, σ_r'', $L\Lambda_r a$, $\Lambda_r La$, $L^2 a$ are computed at the point (t, x).

LEMMA 10.1. *The remainder ρ in (10.3) satisfies the relations*

$$|\mathbf{E}\rho| = O(h^4), \qquad (10.4)$$

$$\mathbf{E}\rho^2 = O(h^6), \qquad (10.5)$$

$$\left| \mathbf{E}\rho \int_t^{t+h} dw_r(\theta) \right| = O(h^4). \qquad (10.6)$$

This Lemma can be proved similarly as Lemma 8.1. To shorten the exposition we have not listed in detail all assumptions to be imposed upon the coefficients a and σ_r; they are similar to those in Lemma 8.1. Here and below, for shortness reasons we will use, e.g., (10.4) instead of (8.5).

We introduce the notations

$$I_r = \int_t^{t+h} dw_r(\theta), \qquad J_r = \int_t^{t+h} (w_r(\theta) - w_r(t))\, d\theta,$$

$$G_r = \int_t^{t+h} (w_r(\theta) - w_r(t))\, (\theta - t)\, d\theta,$$

$$J_{ir} = \int_t^{t+h} \left(\int_t^{\theta} (w_i(\theta_1) - w_i(t))\, dw_r(\theta_1) \right) d\theta,$$

$$J_{sir} = \int_t^{t+h} \left(\int_t^{\theta} \left(\int_t^{\theta_1} (w_s(\theta_2) - w_s(t))\, dw_i(\theta_2) \right) dw_r(\theta_1) \right) d\theta.$$

LEMMA 10.2. *The following identities hold:*

$$\int_t^{t+h} (\theta - t)\, dw_r(\theta) = hI_r - J_r, \tag{10.7}$$

$$\int_t^{t+h} \left(\int_t^{\theta} (\theta_1 - t)\, dw_r(\theta_1) \right) d\theta = 2G_r - hJ_r, \tag{10.8}$$

$$\int_t^{t+h} \left(\int_t^{\theta} (w_r(\theta_1) - w_r(t))\, d\theta_1 \right) d\theta = hJ_r - G_r, \tag{10.9}$$

$$\int_t^{t+h} \left(\int_t^{\theta} (\theta_1 - t)\, d\theta_1 \right) dw_r(\theta) = \frac{1}{2} h^2 I_r - G_r. \tag{10.10}$$

PROOF. We give a proof of (10.8). We have:

$$d\left(\int_t^{\theta} (\theta_1 - t)\, dw_r(\theta_1) \cdot (\theta - t) \right) = \int_t^{\theta} (\theta_1 - t)\, dw_r(\theta_1) \cdot d\theta + (\theta - t)^2\, dw_r(\theta).$$

Integrating this identity from t to $t+h$ gives

$$\int_t^{t+h} \left(\int_t^{\theta} (\theta_1 - t)\, dw_r(\theta_1) \right) d\theta = h \int_t^{t+h} (\theta - t)\, dw_r(\theta) - \int_t^{t+h} (\theta - t)^2\, dw_r(\theta). \tag{10.11}$$

Further,

$$d\left((\theta - t)^2 (w_r(\theta) - w_r(t)) \right) = 2\, (\theta - t)\, (w_r(\theta) - w_r(t))\, d\theta + (\theta - t)^2\, dw_r(\theta),$$

whence

$$\int_t^{t+h} (\theta - t)^2 \, dw_r(\theta) = h^2 \left(w_r(t+h) - w_r(t)\right) - 2 \int_t^{t+h} (\theta - t) \left(w_r(\theta) - w_r(t)\right) d\theta$$
$$= h^2 I_r - 2G_r. \tag{10.12}$$

Formula (10.7) for the first integral $\int_t^{t+h} (\theta - t) \, dw_r(\theta)$ on the righthand side of (10.11) can be obtained in a similar way. Substituting (10.7) and (10.12) into (10.11) we obtain (10.8). Thus, (10.8) has been proved. The derivation of (10.9) and (10.10) is even simpler. □

By Lemma 10.2, formula (10.3) can be written as

$$\begin{aligned} X(t+h) = x &+ \sum_{r=1}^{q} \sigma_r I_r + ah + \sum_{r=1}^{q} \Lambda_r a J_r + \sum_{r=1}^{q} \sigma'_r \left(hI_r - J_r\right) \\ &+ La\frac{h^2}{2} + \sum_{r=1}^{q}\sum_{i=1}^{q} \Lambda_i \Lambda_r a J_{ir} + \sum_{r=1}^{q} L\Lambda_r a \left(2G_r - hJ_r\right) \\ &+ \sum_{r=1}^{q} \Lambda_r La \left(hJ_r - G_r\right) + \sum_{r=1}^{q} \sigma''_r \left(\frac{1}{2}h^2 I_r - G_r\right) \\ &+ L^2 a\frac{h^3}{6} + \sum_{r=1}^{q}\sum_{i=1}^{q}\sum_{s=1}^{q} \Lambda_s \Lambda_i \Lambda_r a J_{sir} + \rho. \end{aligned} \tag{10.13}$$

LEMMA 10.3. *The following holds:*

$$\mathbf{E} J_{sir} = 0, \qquad \mathbf{E} J_{sir} I_j = 0, \qquad \mathbf{E} J_{sir} I_j I_l = 0. \tag{10.14}$$

PROOF. The first and last identities in (10.14) can be proved by a parity argument. We prove the second identity. Without loss of generality we may put $t = 0$. We introduce the following system of stochastic differential equations:

$$\begin{aligned} dx &= w_s \, dw_i, & x(0) &= 0, \\ dy &= x \, dw_r, & y(0) &= 0, \\ dz &= y \, d\theta, & z(0) &= 0. \end{aligned} \tag{10.15}$$

We have

$$\begin{aligned} z(h) &= \int_0^h y(\theta) \, d\theta = \int_0^h \left(\int_0^\theta x(\theta_1) \, dw_r(\theta_1) \right) d\theta \\ &= \int_0^h \left(\int_0^\theta \left(\int_0^{\theta_1} w_s(\theta_2) \, dw_i(\theta_2) \right) dw_r(\theta_1) \right) d\theta = I_{sir}, \end{aligned}$$

$$\mathbf{E} J_{sir} \cdot I_j = \mathbf{E} w_j(h) z(h).$$

We can write

$$dw_j z = z \, dw_j(\theta) + w_j y \, d\theta.$$

Whence

$$\mathbf{E}w_j(h)z(h) = \int_0^h \mathbf{E}(w_j y)\, d\theta. \tag{10.16}$$

Further,

$$dw_j y = y\, dw_j + w_j x\, dw_r + \delta_{jr} x\, d\theta.$$

Therefore

$$\mathbf{E}(w_j y) = \int_0^\theta \delta_{jr} \mathbf{E}x(\theta_1)\, d\theta_1. \tag{10.17}$$

The first equation in (10.15) implies $\mathbf{E}x \equiv 0$. Therefore (10.17) implies $\mathbf{E}w_j y \equiv 0$, and (10.16) implies $\mathbf{E}(w_j(h)z(h)) = 0$, as desired. □

As in §8 we introduce an auxiliary vector $\widetilde{X}$, equal to the righthand side of (10.13) without the last two terms:

$$\begin{aligned}\widetilde{X} = x &+ \sum_{r=1}^q \sigma_r I_r + ah + \sum_{r=1}^q \Lambda_r a J_r + \sum_{r=1}^q \sigma_r' (hI_r - J_r) + La\frac{h^2}{2} \\ &+ \sum_{r=1}^q \sum_{i=1}^q \Lambda_i \Lambda_r a J_{ir} + \sum_{r=1}^q L\Lambda_r a\,(2G_r - hJ_r) + \sum_{r=1}^q \Lambda_r La\,(hJ_r - G_r) \\ &+ \sum_{r=1}^q \sigma_r'' \left(\frac{1}{2}h^2 I_r - G_r\right) + L^2 a \frac{h^3}{6}.\end{aligned} \tag{10.18}$$

LEMMA 10.4. *The following relations hold:*

$$\left| \mathbf{E}\left(\prod_{j=1}^s \Delta^{i_j} - \prod_{j=1}^s \widetilde{\Delta}^{i_j} \right) \right| = O(h^4), \qquad s = 1, \ldots, 7. \tag{10.19}$$

The proof of this Lemma is based on Lemma 10.1 and Lemma 10.3, and is not essentially different from the proof of Lemma 8.3.

10.2. Construction of a one-step approximation of order of accuracy four, and of a method of order three. We construct the *one-step approximation* $\overline{X}$ on the basis of $\widetilde{X}$ as follows:

$$\begin{aligned}\overline{X} = x &+ \sum_{r=1}^q \sigma_r \xi_r h^{1/2} + ah + \sum_{r=1}^q \Lambda_r a \eta_r h^{3/2} + \sum_{r=1}^q \sigma_r' (\xi_r - \eta_r)\, h^{3/2} + La\frac{h^2}{2} \\ &+ \sum_{r=1}^q \sum_{i=1}^q \Lambda_i \Lambda_r a \eta_{ir} h^2 + \sum_{r=1}^q L\Lambda_r a\,(2\mu_r - \eta_r)\, h^{5/2} + \sum_{r=1}^q \Lambda_r La\,(\eta_r - \mu_r)\, h^{5/2} \\ &+ \sum_{r=1}^q \sigma_r'' \left(\frac{1}{2}\xi_r - \mu_r\right) h^{5/2} + L^2 a \frac{h^3}{6}.\end{aligned} \tag{10.20}$$

To construct a method of order of accuracy three (the order of accuracy of such a method equals 4 at each step) we have to get convinced of the relations

$$\left|\mathbf{E}\left(\prod_{j=1}^{s}\widetilde{\Delta}^{i_j}-\prod_{j=1}^{s}\overline{\Delta}^{i_j}\right)\right|=O(h^4),\qquad s=1,\dots,7. \tag{10.21}$$

In fact, by Lemma 10.4 in this case we will also have

$$\left|\mathbf{E}\left(\prod_{j=1}^{s}\Delta^{i_j}-\prod_{j=1}^{s}\overline{\Delta}^{i_j}\right)\right|=O(h^4),\qquad s=1,\dots,7, \tag{10.22}$$

and, by Theorem 9.1, a method based on the one-step approximation $\overline{X}$ with the properties (10.22) and (10.23) will have order of accuracy three. Of course, the above said is valid under the standard assumptions regarding the coefficients of the system (10.1) which, in particular for the approximation (10.20), ensure the relation

$$\mathbf{E}\prod_{j=1}^{s}\left|\overline{\Delta}^{i_j}\right|=O(h^4),\qquad s=8. \tag{10.23}$$

So, the chase starts with the relations (10.21).

LEMMA 10.5. *The following seven groups of identities ensure the relations* (10.21)*:*

$$\begin{aligned}
&\mathbf{E}\xi_r h^{1/2}=\mathbf{E}I_r=0, &&\mathbf{E}\eta_r h^{3/2}=\mathbf{E}J_r=0,\\
&\mathbf{E}\eta_{ir}h^2=\mathbf{E}J_{ir}=0, &&\mathbf{E}\mu_r h^{5/2}=\mathbf{E}G_r=0;
\end{aligned} \tag{10.24}$$

$$\begin{aligned}
&\mathbf{E}\xi_i\xi_r h=\mathbf{E}I_iI_r=\delta_{ir}h, &&\mathbf{E}\xi_r\eta_j h^2=\mathbf{E}I_rJ_j=\delta_{rj}\frac{h^2}{2},\\
&\mathbf{E}\xi_i\eta_{jr}h^{5/2}=\mathbf{E}I_iJ_{jr}=0, &&\mathbf{E}\xi_i\mu_r h^3=\mathbf{E}I_iG_r=\delta_{ir}\frac{h^3}{3},\\
&\mathbf{E}\eta_i\eta_j h^3=\mathbf{E}J_iJ_j=\delta_{ij}\frac{h^3}{3}, &&\mathbf{E}\eta_i\eta_{jr}h^{7/2}=\mathbf{E}J_iJ_{jr}=0;
\end{aligned} \tag{10.25}$$

$$\begin{aligned}
\mathbf{E}\xi_i\xi_r\eta_{js}h^3&=\mathbf{E}I_iI_rJ_{js}\\
&=\begin{cases}\frac{h^3}{6} & \text{if } j\neq s \text{ and either } i=j,\ r=s \text{ or } i=s,\ j=r,\\ \frac{h^3}{3} & \text{if } i=r=j=s,\\ 0 & \text{otherwise,}\end{cases}\\
\mathbf{E}\xi_i\xi_r\mu_j h^{7/2}&=\mathbf{E}I_iI_rG_j=0,\qquad \mathbf{E}\xi_i\eta_j\eta_r h^{7/2}=\mathbf{E}I_iJ_jJ_r=0;
\end{aligned} \tag{10.26}$$

$$\mathbf{E}\xi_i\xi_r\xi_j\xi_s = \mathbf{E}I_iI_rI_jI_s = \begin{cases} h^2 & \textit{if } \{i,r,j,s\} \textit{ consists of two pairs of equal numbers,} \\ 3h^2 & \textit{if } i=r=j=s, \\ 0 & \textit{otherwise,} \end{cases}$$

$$\mathbf{E}\xi_i\xi_r\xi_j\eta_s h^3 = \mathbf{E}I_iI_rI_jJ_s = \begin{cases} \frac{h^3}{2} & \textit{if } \{i,r,j,s\} \textit{ consists of two pairs of equal numbers,} \\ \frac{3h^2}{2} & \textit{if } i=r=j=s, \\ 0 & \textit{otherwise,} \end{cases}$$

$$\mathbf{E}\xi_i\xi_r\xi_j\eta_{sl}h^{7/2} = \mathbf{E}I_iI_rI_jJ_{sl} = 0; \tag{10.27}$$

$$\mathbf{E}\xi_i\xi_r\xi_j\xi_s\xi_l h^{5/2} = \mathbf{E}I_iI_rI_jI_sI_l = 0, \qquad \mathbf{E}\xi_i\xi_r\xi_j\xi_s\eta_l h^{7/2} = \mathbf{E}I_iI_rI_jI_sJ_l = 0; \tag{10.28}$$

in the following identities we assume, without loss of generality, that $i_1 \le i_2 \le i_3 \le i_4 \le i_5 \le i_6$:

$$\mathbf{E}\prod_{j=1}^{6}\xi_{i_j}h^3 = \mathbf{E}\prod_{j=1}^{6}I_{i_j} = \begin{cases} h^3, & i_1=i_2<i_3=i_4<i_5=i_6, \\ 3h^3, & i_1=i_2<i_3=i_4=i_5=i_6, \\ 3h^3, & i_1=i_2=i_3=i_4<i_5=i_6, \\ 15h^3, & i_1=i_2=i_3=i_4=i_5=i_6, \\ 0 & \textit{otherwise;} \end{cases} \tag{10.29}$$

$$\mathbf{E}\prod_{j=1}^{7}\xi_{i_j}h^{7/2} = \mathbf{E}\prod_{j=1}^{7}I_{i_j} = 0. \tag{10.30}$$

PROOF. The proof of this Lemma in many respects repeats the proof of Lemma 8.4. Here we will consider in some detail only the proof of the identity for $I_iI_rI_{js}$ in (10.26). Without loss of generality we may put $t = 0$. We introduce the equations

$$dx = w_j\,dw_s(\theta), \qquad x(0) = 0,$$
$$dy = x\,d\theta, \qquad y(0) = 0.$$

Then

$$y(h) = \int_0^h x(\theta)\,d\theta = \int_0^h \left(\int_0^\theta w_j(\theta_1)\,dw_s(\theta_1)\right) d\theta = J_{js}(h),$$
$$\mathbf{E}I_iI_rI_{js} = \mathbf{E}\left(w_i(h)w_r(h)y\right).$$

We have

$$d\left(w_iw_ry\right) = w_ry\,dw_i + w_iy\,dw_r + w_iw_rx\,d\theta + y\delta_{ir}\,d\theta.$$

Hence

$$d\mathbf{E}\,(w_i w_r y) = \mathbf{E}\,(w_i w_r x)\; d\theta, \tag{10.31}$$

since, obviously, $\mathbf{E}y$ vanishes. We turn to the computation of $\mathbf{E}w_i w_r x$. We have:

$$d\,(w_i w_r x) = w_r x\, dw_i + w_i x\, dw_r + w_i w_r w_j\, dw_s + x\delta_{ir}\, d\theta + w_i w_j \delta_{rs}\, d\theta + w_r w_j \delta_{is}\, d\theta.$$

Whence,

$$d\mathbf{E}\,(w_i w_r x) = \delta_{ir}\mathbf{E}x\, d\theta + \mathbf{E}\,(w_i w_j)\,\delta_{rs}\, d\theta + \mathbf{E}\,(w_r w_j)\,\delta_{is}\, d\theta. \tag{10.32}$$

Since $\mathbf{E}x \equiv 0$, the righthand side of (10.32) does not vanish in three cases only:

First case: $i = j \neq s = r$. We have

$$d\mathbf{E}(w_i w_r x) = \theta\, d\theta, \qquad \mathbf{E}\,(w_i w_r x) = \frac{\theta^2}{2},$$

and (10.31) implies

$$\mathbf{E}I_i I_r I_{js} = \mathbf{E}\,(w_i w_r y) = \frac{h^3}{6}.$$

Second case: $r = j \neq s = i$. This case can be considered in a similar way and leads to the same result.

Third case: $i = j = r = s$. This gives

$$d\mathbf{E}(w_i w_r x) = 2\theta\, d\theta, \qquad \mathbf{E}I_i I_r J_{js} = \frac{h^3}{3}.$$

So, the identity for $\mathbf{E}I_i I_r J_{js}$ in (10.26) has been proved completely. The remaining identities in (10.24)–(10.30) can be proved in a way that is definitely not more complicated. This proves Lemma 10.5. □

For the final construction of the one-step approximation $\overline{X}$ (see (10.20)) it remains to choose the random variables ξ_i, η_r, η_{ir}, μ_r such that the relations (10.24)–(10.30) hold. This can be done by modeling these random variables in various ways. Here we can choose them to be even simpler than in §8: although we are constructing a method of higher order of accuracy, we are doing it for systems of less general form.

We will look for these variables in the following way. Consider symmetric random variables ξ_i, ν_j, ζ_r, $i, j, r = 1, \dots, q$, that are all mutually independent (the condition of symmetry can be replaced by the weaker condition of vanishing of the corresponding odd moments), and put

$$\eta_i = \frac{\xi_i}{2} + \nu_i, \qquad \eta_{ij} = \frac{1}{6}\,(\xi_i\xi_j - \zeta_i\zeta_j), \qquad \mu_i = \frac{\xi_i}{3}. \tag{10.33}$$

Then ξ_i, ν_j, ζ_r have the following moments:

$$\mathbf{E}\xi_i = \mathbf{E}\xi_i^3 = \mathbf{E}\xi_i^5 = \mathbf{E}\xi_i^7 = 0, \qquad \mathbf{E}\nu_j = 0, \qquad \mathbf{E}\zeta_r = 0, \tag{10.34}$$

$$\mathbf{E}\xi_i^2 = 1, \qquad \mathbf{E}\xi_i^4 = 3, \qquad \mathbf{E}\xi_i^6 = 15, \qquad \mathbf{E}\nu_j^2 = \frac{1}{12}, \qquad \mathbf{E}\zeta_i^2 = 1. \tag{10.35}$$

LEMMA 10.6. *Suppose that ξ_i, ν_j, ζ_r are independent random variables with moments satisfying* (10.34)–(10.35). *Then the variables* (10.33) *satisfy the relations* (10.26)–(10.30).

The proof of this Lemma consists of a simple verification of the relations (10.26)–(10.30).

For the identities (10.34)–(10.35) to be satisfied, the random variables ν_j and ζ_r are simplest of all modeled by the laws $\mathbf{P}(\nu = -1/\sqrt{12}) = \mathbf{P}(\nu = 1/\sqrt{12}) = 1/2$, $\mathbf{P}(\zeta = -1) = \mathbf{P}(\zeta = 1) = 1/2$, while ξ_i can be modeled by the law $N(0,1)$. However, for ξ_i we can also choose a simpler law. For example, $\mathbf{P}(\xi = 0) = 1/3$, $\mathbf{P}(\xi = -1) = \mathbf{P}(\xi = 1) = 3/10$, $\mathbf{P}(\xi = -\sqrt{6}) = \mathbf{P}(\xi = \sqrt{6}) = 1/30$.

Since (10.22)–(10.23) hold, the one-step approximation (10.20) with random variables (10.33) has order of accuracy four.

We sum up the results obtained in the following Theorem.

THEOREM 10.1. *Suppose the coefficient $a(t,x)$ in the system* (10.1) *satisfies the Lipschitz condition*

$$|a(t,x) - a(t,y)| \le K\,|x-y|\,.$$

Suppose that $a(t,x)$, together with its partial derivatives up to sufficiently high order (at least up to order seven, inclusively) belongs to F, and suppose that the coefficients $\sigma_r(t)$ are three times continuously differentiable with respect to $t \in [t_0, t_0+T]$. Suppose that the functions a, $\Lambda_r a$, La, $\Lambda_i\Lambda_r a$, $L\Lambda_r a$, $\Lambda_r La$, and $L^2 a$ grow at most linearly as $|x|$ goes to infinity. Then the method

$$\begin{aligned}
X_{k+1} = X_k &+ \sum_{r=1}^{q} \sigma_r(t_k)\xi_{rk}h^{1/2} + a_k h + \sum_{r=1}^{q} (\Lambda_r a)_k \left(\frac{\xi_{rk}}{2} + \nu_{rk}\right) h^{3/2} \\
&+ \sum_{r=1}^{q} \sigma_r'(t_k)\left(\frac{\xi_{rk}}{2} - \nu_{rk}\right) h^{3/2} + (La)_k \frac{h^2}{2} \\
&+ \frac{1}{6}\sum_{r=1}^{q}\sum_{i=1}^{q} (\Lambda_i\Lambda_r a)_k \left(\xi_{ik}\xi_{rk} - \zeta_{ik}\zeta_{rk}\right) h^2 \\
&+ \sum_{r=1}^{q} (L\Lambda_r a)_k \left(\frac{1}{6}\xi_{rk} - \nu_{rk}\right) h^{5/2} + \sum_{r=1}^{q} (\Lambda_r La)_k \left(\frac{1}{6}\xi_{rk} + \nu_{rk}\right) h^{5/2} \\
&+ \frac{1}{6}\sum_{r=1}^{q} \sigma_r''(t_k)\xi_{rk}h^{5/2} + \left(L^2 a\right)_k \frac{h^3}{6},
\end{aligned} \tag{10.36}$$

where the sets of random variables ξ_{ik}, ν_{rk}, ζ_{jk} for various k are independent, and for each k are modeled as independent variables such that the relations (10.34)–(10.35) *hold, is a method of order of accuracy three in the sense of weak approximation* (*relation* (9.5) *with $p = 3$ holds for all functions f belonging, together with their partial derivatives up to order eight, inclusively, to the class F*).

Thanks to the Lemmas proved in this Subsection, the proof of this Theorem follows immediately from Lemma 9.1 and Theorem 9.1.

11. An implicit method

In this Section we construct a two-parameter family of methods of order of accuracy two in the sense of weak approximations for systems of general form (8.1). As in the case of deterministic systems and mean-square approximation, such methods are necessary for integrating stiff systems. The notions of region of stability and of A-stability are completely preserved here, since they are related only with the choice of a test equation and with properties of the difference equation corresponding to one or other method. We will use the same assumptions and notations as in §8 and §9.

The following formula holds (cf. §2.2):

$$a(t+h, X(t+h)) = a + \sum_{r=1}^{q} \Lambda_r a \int_t^{t+h} dw_r(\theta) + La \cdot h + \rho_1, \tag{11.1}$$

where

$$\begin{aligned}\rho_1 = {} & \sum_{r=1}^{q}\sum_{i=1}^{q} \Lambda_i \Lambda_r a \int_t^{t+h} \left(\int_t^{\theta} dw_i(\theta_1) \right) dw_r(\theta) \\ & + \sum_{r=1}^{q}\sum_{i=1}^{q}\sum_{s=1}^{q} \int_t^{t+h} \left(\int_t^{\theta} \left(\int_t^{\theta_1} \Lambda_s \Lambda_i \Lambda_r a(\theta_2, X(\theta_2))\, dw_s(\theta_2) \right) dw_i(\theta_1) \right) dw_r(\theta) \\ & + \sum_{r=1}^{q}\sum_{i=1}^{q} \int_t^{t+h} \left(\int_t^{\theta} \left(\int_t^{\theta_1} L\Lambda_i \Lambda_r a(\theta_2, X(\theta_2))\, d\theta_2 \right) dw_i(\theta_1) \right) dw_r(\theta) \\ & + \sum_{r=1}^{q} \int_t^{t+h} \left(\int_t^{\theta} L\Lambda_r a(\theta_1, X(\theta_1))\, d\theta_1 \right) dw_r(\theta) \\ & + \sum_{r=1}^{q} \int_t^{t+h} \left(\int_t^{\theta} \Lambda_r La(\theta_1, X(\theta_1))\, dw_r(\theta_1) \right) d\theta. \end{aligned} \tag{11.2}$$

As in Lemma 8.1 we can show that

$$|\mathbf{E}\rho_1| \le K(x)h^2, \qquad K(x) \in F, \tag{11.3}$$

$$\mathbf{E}\rho_1^2 \le K(x)h^2, \qquad K(x) \in F, \tag{11.4}$$

$$\left| \mathbf{E}\rho_1 \int_t^{t+h} dw_r(\theta) \right| \le K(x)h^2, \qquad K(x) \in F. \tag{11.5}$$

Further,

$$La(t+h, X(t+h)) = La + \rho_2, \tag{11.6}$$

where

$$\rho_2 = \sum_{r=1}^{q} \int_t^{t+h} \Lambda_r La(\theta, X(\theta))\, dw_r(\theta) + \int_t^{t+h} L^2 a(\theta, X(\theta))\, d\theta. \tag{11.7}$$

We can readily prove that

$$|\mathbf{E}\rho_2| \le K(x)h, \qquad K(x) \in F, \tag{11.8}$$

$$\mathbf{E}\rho_2^2 \le K(x)h, \qquad K(x) \in F, \tag{11.9}$$

$$\left|\mathbf{E}\rho_2 \int_t^{t+h} dw_r(\theta)\right| \le K(x)h, \qquad K(x) \in F. \tag{11.10}$$

Using the relations (8.3), (11.1), and (11.6), it is easy to obtain the following formula, involving the arbitrary constants α, β:

$$\begin{aligned} X(t+h) &= x + \sum_{r=1}^{q} \sigma_r I_r + \alpha a h + (1-\alpha) a(t+h, X(t+h))h \\ &+ \sum_{r=1}^{q}\sum_{i=1}^{q} \Lambda_i \sigma_r I_{ir} + \sum_{r=1}^{q} \left(L\sigma_r - (1-\alpha)\Lambda_r a\right) I_r h + \sum_{r=1}^{q} \left(\Lambda_r a - L\sigma_r\right) J_r \\ &+ \beta(2\alpha - 1) La \cdot \frac{h^2}{2} + (1-\beta)(2\alpha-1) La(t+h, X(t+h)) \cdot \frac{h^2}{2} \\ &+ \sum_{r=1}^{q}\sum_{i=1}^{q}\sum_{s=1}^{q} \Lambda_s \Lambda_i \sigma_r I_{sir} + \rho - (1-\alpha)\rho_1 h - (1-\beta)(2\alpha-1)\rho_2 h^2. \end{aligned} \tag{11.11}$$

Introduce the two-parameter family of implicit methods determined by the formula

$$\begin{aligned} \overline{X} &= x + \sum_{r=1}^{q} \sigma_r \xi_r h^{1/2} + \alpha a h + (1-\alpha) a(t+h, \overline{X}) h \\ &+ \sum_{r=1}^{q}\sum_{i=1}^{q} \Lambda_i \sigma_r \xi_{ir} h + \sum_{r=1}^{q} \left(\frac{1}{2} L\sigma_r + \frac{2\alpha - 1}{2} \Lambda_r a\right) \xi_r h^{3/2} \\ &+ \beta(2\alpha-1) La \cdot \frac{h^2}{2} + (1-\beta)(2\alpha-1) La(t+h, \overline{X}) \cdot \frac{h^2}{2}, \end{aligned} \tag{11.12}$$

where ξ_r, ξ_{ir} can be modeled as in §8.

We show that (under certain natural assumptions) the method (11.12) has order of accuracy two. To this end we consider the equation

$$X - (1-\alpha) a(t+h, X)h - (1-\beta)(2\alpha-1) La(t+h, X)\frac{h^2}{2} = Z. \tag{11.13}$$

We assume that for sufficiently small h and all Z this equation can be solved for X:

$$X = \varphi(t+h, Z). \tag{11.14}$$

Introduce vectors Y, $\overline{Y}$:

$$\begin{aligned} Y = x &+ \sum_{r=1}^{q} \sigma_r I_r + \alpha a h + \sum_{r=1}^{q}\sum_{i=1}^{q} \Lambda_i \sigma_r I_{ir} \\ &+ \sum_{r=1}^{q} \left(L\sigma_r - (1-\alpha)\Lambda_r a\right) I_r h + \sum_{r=1}^{q} \left(\Lambda_r a - L\sigma_r\right) J_r \\ &+ \beta(2\alpha - 1) La \cdot \frac{h^2}{2} \\ &+ \sum_{r=1}^{q}\sum_{i=1}^{q}\sum_{s=1}^{q} \Lambda_s \Lambda_i \sigma_r I_{sir} + \rho - (1-\alpha)\rho_1 h - (1-\beta)(2\alpha-1)\rho_2 h^2, \end{aligned} \tag{11.15}$$

$$\begin{aligned} \overline{Y} = x &+ \sum_{r=1}^{q} \sigma_r \xi_r h^{1/2} + \alpha a h + \sum_{r=1}^{q}\sum_{i=1}^{q} \Lambda_i \sigma_r \xi_{ir} h \\ &+ \sum_{r=1}^{q} \left(\frac{1}{2} L\sigma_r + \frac{2\alpha - 1}{2} \Lambda_r a\right) \xi_r h^{3/2} + \beta(2\alpha - 1) La \cdot \frac{h^2}{2}. \end{aligned} \tag{11.16}$$

Then by (11.11), (11.13), and (11.15):

$$X(t+h) = \varphi(t+h, Y),$$

and by (11.12), (11.13), and (11.16):

$$\overline{X} = \overline{X}(t+h) = \varphi(t+h, \overline{Y}).$$

Assume that the function $\varphi(t+h, y)$ has partial derivatives with respect to y up to order six, inclusively, and that these, together with φ, belong to F. For $s = 1, \dots, 5$ we write

$$\begin{aligned} \left| \mathbf{E} \left(\prod_{j=1}^{s} \Delta^{i_j} - \prod_{j=1}^{s} \overline{\Delta}^{i_j} \right) \right| &= \left| \mathbf{E} \left(\prod_{j=1}^{s} \left(X^{i_j} - x^{i_j} \right) - \prod_{j=1}^{s} \left(\overline{X}^{i_j} - x^{i_j} \right) \right) \right| \\ &= \left| \mathbf{E} \left(\prod_{j=1}^{s} \left(\varphi^{i_j}(t+h, Y) - x^{i_j} \right) - \prod_{j=1}^{s} \left(\varphi^{i_j}(t+h, \overline{Y}) - x^{i_j} \right) \right) \right|. \end{aligned} \tag{11.17}$$

The righthand side of (11.17) is $O(h^3)$ if the differences of all moments have this property (see the proof of Theorem 8.1):

$$\left| \mathbf{E} \left(\prod_{k=1}^{s} \left(Y^{r_k} - x^{r_k} \right) - \prod_{k=1}^{s} \left(\overline{Y}^{r_k} - x^{r_k} \right) \right) \right| \le K(x) h^3, \qquad s = 1, \dots, 5, \quad K(x) \in F, \tag{11.18}$$

and if the following relations hold:

$$\mathbf{E} \prod_{k=1}^{s} \left| Y^{r_k} - x^{r_k} \right| \le K(x) h^3, \qquad s = 6, \quad K(x) \in F, \tag{11.19}$$

$$\mathbf{E} \prod_{k=1}^{s} \left| \overline{Y}^{r_k} - x^{r_k} \right| \le K(x) h^3, \qquad s = 6, \quad K(x) \in F. \tag{11.20}$$

Taking into account the properties (11.3)–(11.5) and (11.8)–(11.10) of the remainders ρ_1 and ρ_2, the relations (11.18)–(11.20) can be proved similarly as in §8 we have proved the analogous relations for the differences $X^{i_j} - x^{i_j}$ and $\overline{X}^{i_j} - x^{i_j}$. So, we have proved that $\overline{X}$, which is implicitly defined by (11.12), satisfies

$$\left|\mathbf{E}\left(\prod_{j=1}^{s} \Delta^{i_j} - \prod_{j=1}^{s} \overline{\Delta}^{i_j}\right)\right| \leq K(x)h^3, \qquad s = 1, \ldots, 5, \quad K(x) \in F. \tag{11.21}$$

We will prove the inequality

$$\mathbf{E}\prod_{j=1}^{s} \left|\overline{\Delta}^{i_j}\right| \leq K(x)h^3, \qquad s = 6, \quad K(x) \in F. \tag{11.22}$$

In fact, the solvability of (11.12) for $\overline{X}$ in the form $\overline{X} = \varphi(t+h, \overline{Y})$ with $\varphi \in F$, implies the existence of all sufficiently high moments of $\overline{X}$, if only ξ_r and ξ_{ir} (which participate in the formula for $\overline{Y}$) have sufficiently high moments. Further, since $a \in F$ and $La \in F$, this implies the existence of moments (up to order six, inclusively) for $a(t+h, \overline{X})$ and $La(t+h, \overline{X})$. Now (11.22) follows immediately from (11.12). Finally, assume that $\varphi(t+h, \overline{Y})$ grows at most linearly as $|x|$ goes to infinity. Then the subsequent application of Lemma 9.1 and Theorem 9.1 leads to a result which we state as a Theorem:

THEOREM 11.1. *Suppose that for sufficiently small h the relation* (11.13) *is solvable for Z:* $X = \phi(t+h, Z)$. *Suppose that the function* $\varphi(t+h, y)$ *has partial derivatives with respect to y up to order six, inclusively, that belong, together with* φ, *to F. Finally, assume that the superposition* $\varphi(t+h, \overline{Y})$, *with* $\overline{Y}$ *defined by* (11.16), *grows at most linearly as* $|x| \to \infty$. *Then the implicit method based on* (11.12) *has order of accuracy two in the sense of weak approximation.*

12. Reducing the error of the Monte-Carlo method

If we compute $\mathbf{E}f(X(t_0+T))$ by the Monte-Carlo method, using an approximate method for integrating the system

$$dX = a(t, X)\,dt + \sum_{r=1}^{q} \sigma_r(t, X)\,dw_r(t) \tag{12.1}$$

to find $X(t_0+T)$, two errors arise. One of these is the *numerical integration error*:

$$\mathbf{E}f(X(t_0+T)) = \mathbf{E}f(\overline{X}(t_0+T)) + O(h^p).$$

The other is the *error* of the Monte-Carlo method:

$$\mathbf{E}f(\overline{X}(t_0+T)) = \frac{1}{N}\sum_{i=1}^{N} f(\overline{X}^{(i)}(t_0+T)) \pm c\frac{(\mathbf{V}f(\overline{X}(t_0+T)))^{1/2}}{N^{1/2}},$$

where, e.g., the values $c = 1, 2, 3$ correspond to the fiducial probabilities 0.68, 0.95, 0.997, respectively.

Since $\mathbf{V}f(\overline{X}(t_0+T))$ is close to $\mathbf{V}f(X(t_0+T))$, we may assume that the error of the Monte-Carlo method can be bounded by $(\mathbf{V}f(X(t_0+T))/N)^{1/2}$. If $\mathbf{V}f(X(t_0+T))$

is large, then in order to achieve a satisfactory precision we have to take into account a very large number of trajectories. If it would be possible to construct instead of $f(X(t_0+T))$ a variable Z such that $\mathbf{E}Z = \mathbf{E}f(X(t_0+T))$ but with $\mathbf{V}Z$ substantially smaller than $\mathbf{V}f(X(t_0+T))$, then the modeling of Z instead of $f(X(t_0+T))$ would make it possible to obtain more precise results with the same amount of computations.

Consider, instead of (12.1), the system

$$dX = a(t,X)\,dt - \sum_{r=1}^{q} \mu_r(t,X)\sigma_r(t,X)\,dt + \sum_{r=1}^{q} \sigma_r(t,X)\,dw_r,$$

$$dY = \sum_{r=1}^{q} \mu_r(t,X)Y\,dw_r, \tag{12.2}$$

where μ_r and Y are scalars.

By *Girsanov's Theorem*, for any μ_r:

$$y\mathbf{E}f(X_{s,x}(t_0+T))|_{(12.1)} = \mathbf{E}\left(Y_{s,x,y}(t_0+T)f(X_{s,x}(t_0+T))\right)|_{(12.2)}.$$

Putting $Z = Y_{s,x,y}(t_0+T)f(X_{s,x}(t_0+T))$, we see that $\mathbf{E}Z$ does not depend on the choice of the μ_r, while for $y=1$ it equals the looked for quantity. At the same time, $\mathbf{V}Z$ does depend on the μ_r. In relation with the above said, it is natural to regard $\mu_1,\dots,\mu_q$ as controls and to choose them by the condition that the variance $\mathbf{V}Z = \mathbf{E}Z^2 - (\mathbf{E}Z)^2$ be minimal. Since $\mathbf{E}Z$ is independent of $\mu_1,\dots,\mu_q$, this choice reduces to solving the following problem from optimal control theory: it is required to choose controls $\mu_1,\dots,\mu_q$ constituting a minimum of the functional

$$I = \mathbf{E}\left(Y_{s,x,y}^2(t_0+T)f^2(X_{s,x}(t_0+T))\right)$$

given (12.2).

The function $u(s,x) = \mathbf{E}f(X_{s,x}(t_0+T))|_{(12.1)}$ satisfies the equation

$$Lu \equiv \frac{\partial u}{\partial s} + \sum_{i=1}^{n} a^i\frac{\partial u}{\partial x^i} + \frac{1}{2}\sum_{r=1}^{q}\sum_{i=1}^{n}\sum_{j=1}^{n} \sigma_r^i\sigma_r^j\frac{\partial^2 u}{\partial x^i\partial x^j} = 0 \tag{12.3}$$

with the condition

$$u(t_0+T,x) = f(x) \tag{12.4}$$

at the righthand end t_0+T playing the role of initial condition.

Introduce the function

$$v(s,x)y^2 = \min_{\mu_1,\dots,\mu_q} I = \min_{\mu_1,\dots,\mu_q} \mathbf{E}\left(Y_{s,x,y}^2(t_0+T)f^2(X_{s,x}(t_0+T))\right)$$

(it is clearly homogeneous of order two in y, which is already reflected in the notation). We write down the *Bellman equation* for this function:

$$\min_{\mu_1,\dots,\mu_q}\left(Lvy^2 + \sum_{r=1}^{q}\left(\sigma_r, \frac{\partial v}{\partial x}\right)\mu_r y^2 + v\sum_{r=1}^{q}\mu_r^2 y^2\right) = 0. \tag{12.5}$$

The minimisation condition in (12.5) implies (if $v \neq 0$):

$$\mu_r = -\frac{1}{2v}\left(\sigma_r, \frac{\partial v}{\partial x}\right). \tag{12.6}$$

Thus, v satisfies the equation

$$Lv - \frac{1}{4v}\sum_{r=1}^{q}\left(\sigma_r, \frac{\partial v}{\partial x}\right)^2 = 0. \tag{12.7}$$

Moreover, clearly

$$v(t_0 + T, x) = f^2(x). \tag{12.8}$$

Let $f > 0$. Then $v > 0$. After some simple computations we are readily convinced of the fact that $\sqrt{v}$ is a solution of the problem (12.3)–(12.4). Thus, $v = u^2$. By (12.6), this implies

$$\mu_r = -\frac{1}{u}\left(\sigma_r, \frac{\partial u}{\partial x}\right). \tag{12.9}$$

Further, if we write the relation $v = u^2$ in the form

$$\mathbf{E}Z^2 = \pm(\mathbf{E}Z)^2,$$

then we find that, with μ_r as in (12.9), $\mathbf{V}Z = 0$, i.e. the variable $Y_{s,x,y}(t_0+T)f(X_{s,x}(t_0+T))$, computed from (12.2) with (12.9) taken into account, is deterministic.

Of course, the controls μ_r, $r = 1, \dots, q$, cannot be constructed without knowing the function u. Nevertheless, the result obtained establishes that, in principle, it is possible to arbitrarily reduce the variance $\mathbf{V}Z$ by conveniently choosing the functions μ_i.

Note that the reasoning above is not completely rigorous. However, using the results of it, it is not difficult to state and prove the following Theorem.

THEOREM 12.1. *Let $f > 0$ and suppose there is a solution $u > 0$ of the problem (12.3)–(12.4). Suppose there is a solution of the system (12.2), with μ_r as in (12.9), for $s \le t \le t_0 + T$ for any $t_0 \le s < t_0 + T$ and $x \in \mathbf{R}^n$. Then $Z = Y_{s,x,y}(t_0 + T)f(X_{s,x}(t_0 + T))$, computed according to (12.2) with (12.9) taken into account, is a deterministic variable.*

PROOF. Let $u > 0$ be a solution of (12.3)–(12.4), and let μ_r in (12.2) be such that there is a solution of the system (12.2). Using Itô's formula, we can compute that (taking into account that $Lu = 0$):

$$\begin{aligned} d\left(u(t, X_{s,x}(t)) \cdot Y_{s,x,y}(t)\right) &= Lu \cdot Y\,dt - \sum_{r=1}^{q}\mu_r\left(\sigma_r, \frac{\partial u}{\partial x}\right)Y\,dt \\ &+ \sum_{r=1}^{q}\left(\sigma_r, \frac{\partial u}{\partial x}\right)Y\,dw_r(t) + u\sum_{r=1}^{q}\mu_r Y\,dw_r(t) \\ &+ \sum_{r=1}^{q}\left(\sigma_r, \frac{\partial u}{\partial x}\right)\mu_r Y\,dt + \sum_{r=1}^{q}\left(\left(\sigma_r, \frac{\partial u}{\partial x}\right) + \mu_r u\right)Y\,dw_r(t). \end{aligned}$$

Whence,

$$u(t, X_{s,x}(t)) \cdot Y_{s,x,y}(t) = u(s,x)y + \int_t^s \sum_{r=1}^q \left(\left(\sigma_r, \frac{\partial u}{\partial x} \right) + \mu_r u \right) Y \, dw_r. \quad (12.10)$$

For the μ_r from (12.9) the relation (12.10) reduces to

$$u(t, X_{s,x}(t)) \cdot Y_{s,x,y}(t) = u(s,x)y,$$

i.e. for each t (so, in particular, for $t = t_0 + T$) the quantity $u(t, X_{s,x}(t)) \cdot Y_{s,x,y}(t)$ is deterministic. By (12.4), for $t = t_0 + T$ this quantity is equal to $f(X_{s,x}(t_0 + T))Y_{s,x,y}(t_0 + T)$. $\square$

The results obtained can be used in, e.g., the following situation. Let f be a function close to a function f_0, and let the solution of the problem (12.3)–(12.4) for $f = f_0$ be known and be equal to u_0. Taking μ_r in (12.2) equal to

$$\mu_r = -\frac{1}{u_0} \left(\sigma_r, \frac{\partial u_0}{\partial x} \right),$$

the variance $\mathbf{V}\left(f(x_{s,x}(t_0 + T)) \cdot Y_{s,x,y}(t_0 + T)\right)$, although not zero, will be small. In relation with this, below, in §14, we will give an example.

REMARK 12.1. If the condition $f > 0$ in Theorem 12.1 is not satisfied, but if, e.g., $f > -C$, $C > 0$, then for $f + C$ the solution of the problem (12.3)–(12.4) is $u + C$, and the dependence

$$\mu_r = \frac{1}{u + C} \left(\sigma_r, \frac{\partial u}{\partial x} \right)$$

in (12.2) leads to $Z = Y_{s,x,y}(t_0 + T)\left(f(X_{s,x}(t_0 + T)) + C\right)$ being a deterministic variable (as in Theorem 12.1). If f is neither bounded below nor above, but $f = g - h$ with $g > 0$ and $h > 0$, while for each of the functions g, h the conditions of Theorem 12.1 hold, then to compute $\mathbf{E}f$ Theorem 12.1 can be used for g and h separately.

In conclusion we note that the problem of reducing the variance can be similarly solved also when computing the mathematical expectation of a random variable of far more general form than $f(X_{s,x}(t_0 + T))$. In fact, similar results can be obtained for the computation of $\mathbf{E}\xi_{s,x}$, where

$$\xi_{s,x} = \int_s^\tau \rho(\theta, X_{s,x}(\theta)) \, d\theta + f(\tau, X_{s,x}(\tau)),$$

where ρ and f are known functions, $X_{s,x}(t)$ is the solution of the system (12.1), $\tau = \tau_1 \wedge t_0 + T = \min(\tau_1, t_0 + T)$, τ_1 is the time at which $X_{s,x}(t)$ leaves a given region $D \in \mathbb{R}^n$, $t_0 \le s < t_0 + T$, $x \in D$.

Introducing the function $u(s,x) = \mathbf{E}\xi_{s,x}$, we find

$$\frac{\partial u}{\partial s} + \sum_{i=1}^{n} a^i \frac{\partial u}{\partial x^i} + \frac{1}{2}\sum_{i,j=1}^{n}\sum_{r=1}^{q} \sigma_r^i \sigma_r^j \frac{\partial^2 u}{\partial x^i \partial x^j} + \rho = Lu + \rho = 0, \tag{12.11}$$

$$u(s,x)|_{x\in\Gamma} = f(s,x)|_{x\in\Gamma}, \tag{12.12}$$

$$u(t_0+T,x) = f(t_0+T,x), \quad x \in D, \tag{12.13}$$

where Γ is the boundary of D. Again, instead of (12.1) we consider the system (12.2), and by *Girsanov's Theorem* we find, for any μ_r,

$$\begin{aligned} u(s,x)y &= y\mathbf{E}\xi_{s,x}|_{(12.1)} \\ &= \mathbf{E}\left(\int_s^\tau \rho(\theta, X_{s,x}(\theta))Y_{s,x,y}(\theta)\,d\theta + f(\tau, X_{s,x}(\tau))Y_{s,x,y}(\tau)\right)\Bigg|_{(12.2)} \end{aligned} \tag{12.14}$$

Putting

$$Z(t) = \int_s^t \rho(\theta, X_{s,x}(\theta))Y_{s,x,y}(\theta)\,d\theta + f(t, X_{s,x}(t))Y_{s,x,y}(t), \qquad Z = Z(\tau),$$

we obtain $u(s,x)y = \mathbf{E}Z$, and $\mathbf{E}Z$ does not depend on the choice of the μ_r. At the same time $\mathbf{V}Z$ does not depend on the μ_r. Further, as in the proof of Theorem 12.1, for the function u, which is a solution of the problem (12.11)–(12.13), we obtain instead of (12.10), for $t \le \tau$,

$$\begin{aligned} u(t, X_{s,x}(t)) \cdot Y_{s,x,y}(t) &= u(s,x)y - \int_s^t \rho(\theta, X_{s,x}(\theta))Y_{s,x,y}(\theta)\,d\theta \\ &\quad + \int_s^t \sum_{r=1}^{q}\left(\left(\sigma_r, \frac{\partial u}{\partial x}\right) + \mu_r u\right) \cdot Y\,dw_r. \end{aligned} \tag{12.15}$$

If we take

$$\mu_r = -\frac{1}{u}\left(\sigma_r, \frac{\partial u}{\partial x}\right),$$

then $Z(t)$ and, in particular, Z, is a deterministic quantity; moreover, $Z = u(s,x)y$.

Thus, again we have found μ_r such that $\mathbf{V}Z$ vanishes.

CHAPTER 4

Application of the numerical integration of stochastic equations for the Monte-Carlo computation of Wiener integrals

13. Methods of order of accuracy two for computing Wiener integrals of functionals of integral type

13.1. Statement of the problem. Numerical methods for *Wiener integrals*

$$I = \int\limits_{\mathbf{C}^n} V(x(\cdot))\, d_w x \tag{13.1}$$

are expounded in the books [13], [14], and [45] (see also the references in these books). In this Section we consider a Monte-Carlo method for computing Wiener integrals of functionals of integral type

$$V(x(\cdot)) = \varphi\left(x(T), \int\limits_0^T a(t, x(t))\, dt\right), \qquad x(t) \in \mathbf{R}^n, \tag{13.2}$$

based on the relation between such integrals and stochastic differential equations. Such functionals are often encountered. Let $w(t) = (w^1(t), \dots, w^n(t))$ be an n-dimensional Wiener process. We introduce the system of stochastic differential equations, $t \geq s$,

$$\begin{aligned} dX^1(t) &= dw^1(t), \\ &\dots\dots\dots \\ dX^n(t) &= dw^n(t), \\ dZ(t) &= a(t, X^1(t), \dots, X^n(t))\, dt, \end{aligned} \tag{13.3}$$

with initial conditions

$$X^1(s) = x^1, \dots, X^n(s) = x^n, \quad Z(s) = z. \tag{13.4}$$

We will denote the solution of the system (13.3)–(13.4) by either $X_{s,x}(t)$, $Z_{s,x,z}(t)$, or, if this does not lead to confusion, simply by $X(t)$, $Z(t)$.

The Wiener integral (13.1) of the functional (13.2) is equal to

$$I = \mathbf{E}\varphi(X_{0,0}(T), Z_{0,0,0}(T)). \tag{13.5}$$

According to the Monte-Carlo method, the mathematical expectation $\mathbf{E}\varphi$ can be estimated by the sum

$$I_N = \frac{1}{N}\sum_{m=1}^{N} \varphi(X_{0,0}^{(m)}(T), Z_{0,0,0}^{(m)}(T)), \tag{13.6}$$

where $X^{(m)}(T)$, $Z^{(m)}(T)$, $m = 1, \dots, N$, are independent realisations of the random variables $X(T)$, $Z(T)$.

Thus, finding Wiener integrals of functionals of the form (13.2) leads to the problem of numerically integrating a system of stochastic differential equations (13.3). Of course, for the numerical computation of I it suffices to construct weak approximations of the solution of the system (13.3) and to use the general methods developed in Chapter 3. However, in view of the specificity of the problem under consideration we can construct more efficient methods.

Since (13.3)–(13.4) implies

$$X_{0,0}(t) = w(t), \qquad Z_{0,0,0}(t) = \int_0^t a(\theta, w(\theta))\, d\theta, \tag{13.7}$$

and since at any moment t, $w(t)$ can be modeled exactly, finding an approximation $\overline{Z}_{0,0,0}(T)$ reduces to approximately computing the integral (13.7) for $t = T$. Since there are many efficient quadrature formulas, at first glance this problem does not seem difficult. At the same time we have to keep in mind that quadrature formulas have a high order of accuracy only for integrands that are sufficiently smooth with respect to t. In view of the nonregularity of $w(t)$, the integrand $a(t, w(t))$ in our case will not satisfy the usual conditions of smoothness. Below we will show that the trapezium formula, as applied for computing the integral (13.7), has order of accuracy three at each step, in the sense of weak approximation. In view of the above said, this requires a separate proof, of course.

13.2. Taylor expansions of mathematical expectations. In this and the next Section we will use certain expansions of $\mathbf{E}f(t+h, X(t+h))$ with respect to powers of h which are valid for systems of stochastic differential equations of the general form

$$dX(t) = a(t, X)\, dt + \sum_{r=1}^{q} \sigma_r(t, X)\, dw_r(t). \tag{13.8}$$

By Itô's formula, for $u \geq t$ we have

$$f(u, X(u)) = f(t, X(t)) + \int_t^u Lf(\theta, X(\theta))\, d\theta + \sum_{r=1}^{q} \int_t^u \Lambda_r f(\theta, X(\theta))\, dw_r(\theta), \tag{13.9}$$

where (we recall that)

$$L = \frac{\partial}{\partial \theta} + \left(a, \frac{\partial}{\partial x}\right) + \frac{1}{2}\sum_{r=1}^{q}\left(\sigma_r, \frac{\partial}{\partial x}\right)^2, \qquad \Lambda_r = \left(\sigma_r, \frac{\partial}{\partial x}\right).$$

Applying (13.9) to $Lf(\theta, X(\theta))$ and substituting the expression obtained into $\int_s^t Lf(\theta, X(\theta))\, d\theta$, after a few simple transformations we arrive at

$$f(u, X(u)) = f(t, X(t)) + Lf(t, X(t))(u-t) + \int_t^u (u-\theta)L^2 f(\theta, X(\theta))\, d\theta$$
$$+ \int_t^u \sum_{r=1}^q (\Lambda_r f(\theta, X(\theta)) + (u-\theta)\Lambda_r f(\theta, X(\theta)))\, dw_r(\theta).$$

Proceeding further in this way we find

$$f(u, X(u)) = f(t, X(t)) + Lf(t, X(t))(u-t) + \dots + \frac{1}{m!}L^m f(t, X(t))(u-t)^m$$
$$+ \int_t^u \frac{(u-\theta)^m}{m!} L^{m+1} f(\theta, X(\theta))\, d\theta$$
$$+ \int_t^u \sum_{r=1}^q \Bigg(\Lambda_r f(\theta, X(\theta)) + (u-\theta)\Lambda_r L f(\theta, X(\theta)) + \dots$$
$$+ \frac{(u-\theta)^m}{m!}\Lambda_r L^m f(\theta, X(\theta))\Bigg)\, dw_r(\theta). \tag{13.10}$$

LEMMA 13.1. *Suppose that the following mathematical expectations exist and are continuous with respect to θ:*

$$\mathbf{E}L^k f(\theta, X(\theta)), \qquad k = 0, 1, \dots, m+1,$$
$$\mathbf{E}\left(\Lambda_r L^k f(\theta, X(\theta))\right)^2, \qquad k = 0, \dots, m, \quad r = 1, \dots, q.$$

Then the following formulas hold for $t \le s \le t+h$:

$$\mathbf{E}\left(f(t+h, X_{t,x}(t+h)) \mid \mathcal{F}_s\right) = f(s, X_{t,x}(s)) + (t+h-s)Lf(s, X_{t,x}(s)) + \dots$$
$$+ \frac{(t+h-s)^m}{m!}L^m f(s, X_{t,x}(s))$$
$$+ \int_s^{t+h} \frac{(t+h-\theta)^m}{m!}\mathbf{E}\left(L^{m+1} f(\theta, X_{t,x}(\theta)) \mid \mathcal{F}_s\right) d\theta, \tag{13.11}$$

$$\mathbf{E}f(t+h, X_{t,x}(t+h)) = f(t,x) + hLf(t,x) + \dots$$
$$+ \frac{h^m}{m!}L^m f(t,x) + \int_t^{t+h} \frac{(t+h-\theta)^m}{m!}\mathbf{E}L^{m+1} f(\theta, X(\theta))\, d\theta. \tag{13.12}$$

The proof of this Lemma clearly follows from (13.10).

Formula (13.14) is related to the Taylor expansion of semigroups (see [42, Ch. 11, §2]) and is used here for the same purposes as the expansion of semigroups is used in [25], [26]. It is more convenient than the Taylor expansion of semigroups because, in particular, it is also applicable for, in general, unbounded functions f.

Clearly, in (13.11) and (13.14) the remainders of integral type are $O(h^{m+1})$.

13.3. The trapezium method. We introduce for the system (13.3) the one-step approximation

$$\overline{X}_{t,x}(t+h) = x + \xi h^{1/2},$$
$$\overline{Z}_{t,x,z}(t+h) = z + \frac{h}{2}\left(a(t,x) + a(t+h, \overline{X}_{t,x}(t+h))\right), \tag{13.13}$$

where $\xi = (\xi^1, \dots, \xi^n)$ is an n-dimensional random variable with independent coordinates ξ^i, $i = 1, \dots, n$, such that $\mathbf{E}\xi^i = \mathbf{E}\xi^{i^3} = \mathbf{E}\xi^{i^5} = 0$, $\mathbf{E}\xi^{i^2} = 1$, $\mathbf{E}\xi^{i^4} = 3$. We split the interval $[0,T]$ into N equal parts, with step $h = T/N$: $0 = t_0 < t_1 < \dots < t_N = T$, $t_{k+1} - t_k = h$. Using (13.16) we construct the approximate solution

$$\overline{X}_0 = X_0 = 0, \qquad \overline{X}_{k+1} = \overline{X}_k + \xi_k h^{1/2}, \qquad \overline{Z}_0 = Z_0 = 0,$$
$$\overline{Z}_{k+1} = \overline{Z}_k + \frac{h}{2}\left(a(t_k, \overline{X}_k) + a(t_{k+1}, \overline{X}_{k+1})\right), \qquad k = 0, \dots, N-1, \tag{13.14}$$

where ξ_k, $k = 0, \dots, N-1$ are independent n-dimensional random variables distributed like ξ. We can take, e.g., normally distributed random variables as such ξ_k. However, as in the previous Chapter, in (13.17) we may also use random variables that are more convenient for computing purposes, e.g. with coordinates taking the values 0, $\sqrt{3}$, $-\sqrt{3}$ with probabilities 2/3, 1/6, 1/6.

THEOREM 13.1. *Suppose the function a satisfies a Lipschitz condition. Suppose also that a and φ, together with their partial derivatives with respect to all variables and of order up to six, inclusively, belong to F. Then*

$$\mathbf{E}\varphi(\overline{X}_{0,0}(T), \overline{Z}_{0,0,0}(T)) - \mathbf{E}\varphi(X_{0,0}(T), Z_{0,0,0}(T)) = O(h^2). \tag{13.15}$$

The proof of this Theorem rests on Theorem 9.1. The conditions 1), 3), and 4) of that Theorem (see also Lemma 9.1) are obviously fulfilled. Therefore it remains to verify condition 2). It is not difficult to see that in our case the inequality (9.4) holds. We compute $\mathbf{E}\prod_{j=1}^{s} \Delta^{i_j}$ and $\mathbf{E}\prod_{j=1}^{s} \overline{\Delta}^{i_j}$ for $s = 1, \dots, 5$ to prove that the inequality (9.3) holds.

We introduce $\Delta^i = \Delta X^i = X^i_{t,x}(t+h) - x^i = \xi^i h^{1/2}$ and $\Delta^0 = \Delta Z = Z_{t,x,z}(t+h) - z$. In the sequel we will need the following version of (13.14):

$$f(t+h, X_{t,x}(t+h), Z_{t,x,z}(t+h)) = f(t,x,z) + hLf(t,x,z) + \frac{h^2}{2}L^2 f(t,x,z) + O(h^3), \tag{13.16}$$

where

$$Lf(t,y,v) = \frac{\partial f}{\partial t} + \frac{1}{2}\sum_{j=1}^{n} \frac{\partial^2 f}{\partial y^{j^2}} + a(t,y)\frac{\partial f}{\partial v}.$$

Consider, e.g., the function

$$\psi_i(y,v) = (y^i - x^i)^2 (v - z).$$

Clearly, $\mathbf{E}\Delta^{i^2}\Delta^0 = \mathbf{E}\psi_i(X_{t,x}(t+h), Z_{t,x,z}(t+h))$. We find:

$$L\psi_i(t,y,v) = v - z + a(t,y)\left(y^i - x^i\right)^2,$$
$$L^2\psi_i(t,y,v) = \frac{\partial a}{\partial t}(t,y)\left(y^i - x^i\right)^2 + \frac{1}{2}\sum_{j=1}^{n}\frac{\partial^2 a(t,y)}{\partial y^{j^2}}\left(y^i - x^i\right)^2$$
$$+ 2\frac{\partial a(t,y)}{\partial y^i}\left(y^i - x^i\right) + 2a,$$

whence

$$\psi_i(t,x,z) = 0, \quad L\psi_i(t,x,z) = 0, \quad L^2\psi_i(t,x,z) = 2a(t,x).$$

Therefore, by (13.19) we obtain

$$\mathbf{E}\Delta^{i^2}\Delta^0 = \mathbf{E}\left(\Delta X^i\right)^2\Delta Z = a(t,x)h^2 + O(h^3).$$

Below we write down a list of identities that can be obtained by a similar reasoning or by an even simpler derivation:

$$\begin{aligned}
\mathbf{E}\Delta^{i^2} &= h, \qquad i \neq 0,\\
\mathbf{E}\Delta^0 &= a(t,x)h + \frac{1}{2}\left(\frac{\partial a}{\partial t}(t,x) + \frac{1}{2}\sum_{j=1}^{n}\frac{\partial^2 a}{\partial x^{j^2}}(t,x)\right)h^2 + O(h^3),\\
\mathbf{E}\Delta^i\Delta^0 &= \frac{1}{2}\frac{\partial a}{\partial x^i}(t,x)h^2 + O(h^3), \qquad i \neq 0,\\
\mathbf{E}\Delta^{i^4} &= 3h^2, \qquad i \neq 0,\\
\mathbf{E}\left(\Delta^i\Delta^j\right)^2 &= h^2, \qquad i \neq j, \quad i \neq 0, \quad j \neq 0\\
\mathbf{E}\Delta^{i^2}\Delta^0 &= a(t,x)h^2 + O(h^3), \qquad i \neq 0,\\
\mathbf{E}\Delta^{0^2} &= a^2(t,x)h^2 + O(h^3).
\end{aligned} \tag{13.17}$$

Moreover, all remaining moments $\mathbf{E}\prod_{j=1}^{s}\Delta^{i_j}$, $s = 1,\dots,5$; $i_j = 0,\dots,n$, i.e. not occurring in (13.20), are $O(h^3)$.

To finish the proof of Theorem 13.1 it suffices to show that $\mathbf{E}\prod_{j=1}^{s}\overline{\Delta}^{i_j}$, for $s = 1,\dots,5$, coincides with $\mathbf{E}\prod_{j=1}^{s}\Delta^{i_j}$ up to $O(h^3)$. Write down the identity

$$a(t+h, x+\xi h^{1/2}) = a + \frac{\partial a}{\partial t}h + \sum_{j=1}^{n}\frac{\partial a}{\partial x^j}\xi^j h^{1/2} + \frac{1}{2}\sum_{i,j=1}^{n}\frac{\partial^2 a}{\partial x^i \partial x^j}\xi^i\xi^j h + \rho, \tag{13.18}$$

where a and its derivatives are computed at the point (t,x). The remainder ρ in (13.21) satisfies

$$\mathbf{E}\rho = O(h^2), \qquad \mathbf{E}\rho^2 = O(h^3). \tag{13.19}$$

The relations (13.21), (13.22) can be obtained by introducing the function $\Phi(\lambda) = a(t+\lambda h, x+\lambda\xi h^{1/2})$, $0 \le \lambda \le 1$, and writing down its Taylor expansion:

$$\Phi(1) = \Phi(0) + \Phi'(0) + \frac{1}{2}\Phi''(0) + \frac{1}{6}\Phi'''(0) + \int_0^1 \frac{(1-\lambda)^3}{6}\Phi^{(4)}(\lambda)\,d\lambda.$$

By (13.17),

$$\overline{\Delta}^i = \xi^i h^{1/2}, \qquad \overline{\Delta}^0 = \frac{h}{2}\left(a + a(t+h, x+\xi h^{1/2})\right).$$

It is now not difficult to verify that $\overline{\Delta}^i$, $i = 0, \dots, n$, satisfies the same relations as Δ^i. This proves Theorem 13.1

13.4. The rectangle method and other methods. We can prove that not only the trapezium formula, but any interpolation formula of third order of accuracy with respect to h and intended for computing the integral (13.7) leads to a method of second order of accuracy for computing Wiener integrals. In particular, an application of the rectangle formula gives

$$I = \mathbf{E}\varphi(X_{0,0}(T), Z_{0,0,0}(T)) = \mathbf{E}\varphi\left(w(T), \frac{1}{N}\sum_{k=1}^{N} a(t_{k-1/2}, w(t_{k-1/2}))\right) + O(h^2), \tag{13.20}$$

where $t_{k-1/2} = t_k - h/2$. Note that this formula is substantiated in [46], and also that the Runge–Kutta type method proposed in [54], [59] (see §9.3) leads to the rectangle method when dealing with systems (13.3).

It is clear that $\mathbf{E}\varphi\left(w(T), \frac{1}{N}\sum_{k=1}^{N} a(t_{k-1/2}, w(t_{k-1/2}))\right)$ can be realised in the form $\mathbf{E}\varphi(\overline{X}_{0,0}(T), \overline{Z}_{0,0,0}(T))$ where

$$\overline{X}_0 = X_0 = 0, \qquad \overline{X}_{k+1} = \overline{X}_k + \xi_k h^{1/2}, \qquad \overline{Z}_0 = Z_0 = 0,$$
$$\overline{Z}_{k+1} = \overline{Z}_k + a\left(t_{k+1/2}, \overline{X}_k + \eta\left(\frac{h}{2}\right)^{1/2}\right), \qquad k = 0, \dots, N-1, \tag{13.21}$$

and where the ξ_k, $k = 0, \dots, N-1$, and η are independent n-dimensional random variables whose coordinates, in turn, are independent $N(0,1)$-distributed random variables.

As in the trapezium method, we can also use simpler random variables.

We stress that because of the nonregularity of Brownian trajectories (we spoke of this already in §13.1) separate proofs are required for the result that quadrature formulas for (13.7) under consideration give the same accuracy as in the deterministic case. We have given such proof here for the case of the trapezium formula.

To confirm this we consider the following system of two equations:

$$dX(s) = dw(s), \qquad X(0) = 0,$$
$$dZ(s) = X(s)\,ds, \qquad Z(0) = 0,$$

as well as the function $\varphi(z) = z^2$. Here, $\mathbf{E}z^2(h)$ can be readily computed exactly: $\mathbf{E}Z^2(h) = h^3/3$. We compute the approximation $\overline{Z}(h)$ by Simson's formula:

$$\overline{Z}(h) = \frac{h}{6}\left(X(0) + 4X\left(\frac{h}{2}\right) + X(h)\right) = \frac{h}{6}\left(4X\left(\frac{h}{2}\right) + X(h)\right).$$

We are immediately convinced that $\mathbf{E}\overline{Z}^2(h) = 13\,h^3/36$, and hence that $\mathbf{E}\overline{Z}^2(h) - \mathbf{E}Z^2(h) = O(h^3)$ instead of the expected $O(h^5)$, since to compute the integral (13.7) for $t = h$ we now use Simson's formula, which has order of accuracy five.

In general, we can prove that no one-step approximation of the form (for a function a not depending on t):

$$\begin{aligned}
\overline{X}(t) = x, \qquad &\overline{X}(t+h) = x + \left(\sqrt{\alpha_1}\xi^{(1)} + \sqrt{\alpha_2}\xi^{(2)} + \sqrt{\alpha_3}\xi^{(3)}\right) h^{1/2},\\
\overline{Z}(t) = z, \qquad &\overline{Z}(t+h) = z + p_1 k_1(h) + p_2 k_2(h) + p_3 k_3(h),\\
k_1(h) = hf(x), \qquad &k_2(h) = hf\left(x + \sqrt{\alpha_1}\xi^{(1)} h^{1/2}\right),\\
&k_3(h) = hf\left(x + \left(\sqrt{\alpha_1}\xi^{(1)} + \sqrt{\alpha_2}\xi^{(2)}\right) h^{1/2}\right),
\end{aligned} \tag{13.22}$$

where $\xi^{(1)}, \xi^{(2)}, \xi^{(3)}$ are independent $N(0,1)$-distributed variables, and $0 \le \alpha_1, \alpha_2, \alpha_3 \le 1$, $\alpha_1 + \alpha_2 + \alpha_3 = 1$, are arbitrary real numbers, can achieve order of accuracy four. In (13.25) the expression $\sqrt{\alpha_1}\xi^{(1)} h^{1/2}$ imitates the increment of a Wiener process on $[t, t+\alpha_1 h]$, $\left(\sqrt{\alpha_1}\xi^{(1)} + \sqrt{\alpha_2}\xi^{(2)}\right) h^{1/2}$ imitates such on $[t, t+\alpha_1 h + \alpha_2 h]$, and $\left(\sqrt{\alpha_1}\xi^{(1)} + \sqrt{\alpha_2}\xi^{(2)} + \sqrt{\alpha_3}\xi^{(3)}\right) h^{1/2}$ imitates such on $[t, t+h]$. Thus, the recurrence process (13.25) is constructed similarly to the explicit Runge–Kutta method in the deterministic case, and we can say that there is no ordinary Runge–Kutta method of third order of accuracy for integrating the system (13.3).

13.5. Generalisation of the trapezium formula to Wiener integrals of functionals of general form. Define the step random process $w^h(t)$, $0 \le t \le T$, to be equal to $w(t_k)$ for $t \in [t_k - h/2, t_k + h/2) \cap [0, T]$. Approximation by the *trapezium method* of the integral $\int_0^T a(w(t))\,dt$ is not different from $\int_0^T a(w^h(t))\,dt$. In the previous Section we have proved that the deviation $\mathbf{E}\varphi\left(\int_0^T a(w(t))\,dt\right)$ from the mathematical expectation of such an expression, with integrand replaced by its approximation by the trapezium method, has order $O(h^2)$ if only φ and a are sufficiently smooth functions. There arises the conjecture that this might be true also for functionals $F(x)$ of more general form than $F(x) = \varphi\left(\int_0^T a(x(t))\,dt\right)$. In this Section we show that this conjecture is true for a very large class of functionals. We give all additional assumptions during the course of exposition.

For simplicity of notation we will consider only the case of functionals of one-dimensional functions $x(t)$. We denote by $A = A[0,T]$ the space of right continuous functions on $[0,T]$ that have finite left limits, with norm $\|x\| = \sup_{0\le t\le T} |x(t)|$. We will assume that the functional $F(x)$, $x \in A$, is six times continuously differentiable.

Write down the Taylor expansion:

$$F(x+\delta) = F(x) + DF(x;\delta) + \cdots + \frac{1}{5!}D^5F(x;\delta,\ldots,\delta) + \frac{1}{6!}D^6F(x+\theta\delta;\delta,\ldots,\delta), \qquad 0<\theta<1. \tag{13.23}$$

We will assume that the r-linear functionals $D^rF(x;\delta_1,\ldots,\delta_r)$, $r=1,\ldots,6$, have the form

$$D^rF(x;\delta_1,\ldots,\delta_r) = \int\limits_{[0,T]^r} f^r(x;s_1,\ldots,s_r)\delta_1(s_1)\ldots\delta_r(s_r)\mu_r(ds_1\ldots ds_r), \tag{13.24}$$

where the μ_r are certain fixed finite measures on $[0,T]^r$.

We will not explicitly state the assumptions related with measurability properties and the existence of the mathematical expectations.

Together with $w(t)=w_n(t)$ and $w^h(t)=w_0(t)$ we consider the processes $w_k(t)$, $k=1,\ldots,n-1$, and $\overline{w}_k(t)$, $k=0,\ldots,n-1$, defined by

$$w_k(t) = w(t)\chi_{[t_0,t_k)} + w(t_k)\chi_{[t_k,T]} + \sum_{j=k}^{n-1}\Delta_j w\cdot\chi_{[t_j+h/2,T]}, \tag{13.25}$$

$$\overline{w}_k(t) = w(t)\chi_{[t_0,t_k)} + w(t_k)\chi_{[t_k,T]} + \sum_{j=k+1}^{n-1}\Delta_j w\cdot\chi_{[t_j+h/2,T]}. \tag{13.26}$$

Our aim is to prove (under certain appropriate conditions) the relation

$$\mathbf{E}F(w) - \mathbf{E}F(w^h) = \mathbf{E}F(w_n) - \mathbf{E}F(w_0) = O(h^2). \tag{13.27}$$

Since

$$\mathbf{E}F(w_n) - \mathbf{E}F(w_0) = \sum_{k=0}^{n-1}\left(\mathbf{E}F(w_{k+1}) - \mathbf{E}F(w_k)\right),$$

for this it suffices to prove that

$$\mathbf{E}\left(F(w_{k+1}) - F(w_k)\right) = O(h^3) \tag{13.28}$$

uniformly in k and n. To prove (13.31) we apply Taylor's formula (13.26) to $F(w_k)$ and to $F(w_{k+1})$ with $x=\overline{w}_k$.

Introducing the notations

$$\delta_k^0(t) = w_k(t) - \overline{w}_k(t) = \Delta_k w\cdot\chi_{[t_k+h/2,T]}, \tag{13.29}$$

$$\delta_k^1(t) = w_{k+1}(t) - \overline{w}_k(t) = (w(t)-w(t_k))\cdot\chi_{[t_k,t_{k+1})} + \Delta_k w\cdot\chi_{[t_{k+1},T]}, \tag{13.30}$$

we obtain:

$$\begin{aligned}F(w_{k+i}) = F(\overline{w}_k) + \int\limits_{[t_0,T]} f^1(\overline{w}_k; s)\delta_k^i(s)\mu_1(ds) + \dots \\ + \frac{1}{5!}\int\limits_{[t_0,T]^5} f^5(\overline{w}_k; s_1,\dots,s_5)\delta_k^i(s_1)\dots\delta_k^i(s_5)\mu_5(ds_1\dots ds_5) \\ + \frac{1}{6!}\int\limits_{[t_0,T]^6} f^6(\overline{w}_k + \theta_i\delta_k^i; s_1,\dots,s_6)\delta_k^i(s_1)\dots\delta_k^i(s_6)\mu_6(ds_1\dots ds_6), \\ i = 0, 1. \qquad (13.31)\end{aligned}$$

Formulas (13.29), (13.32), and (13.33) clearly imply that both $\delta_k^0(t)$ and $\delta_k^1(t)$ are independent with $\overline{w}_k(t)$. Therefore the mathematical expectations of the odd terms in (13.34) vanish. If we assume the density $f^6(x; s_1,\dots,s_6)$ to be uniformly bounded, then the mathematical expectation of the last term in (13.34) is $O(h^3)$. As a result,

$$\begin{aligned}&\mathbf{E}\left(F(w_{k+i}) - F(w_k)\right) \\ &= \frac{1}{2!}\mathbf{E}\int\limits_{[t_0,T]^2} f^2(\overline{w}_k; s_1, s_2)\mathbf{E}\left(\delta_k^1(s_1)\delta_k^1(s_2) - \delta_k^0(s_1)\delta_k^0(s_2)\right)\mu_2(ds_1 ds_2) \\ &+ \frac{1}{4!}\mathbf{E}\int\limits_{[t_0,T]^4} f^4(\overline{w}_k; s_1,\dots,s_4)\mathbf{E}\left(\delta_k^1(s_1)\dots\delta_k^1(s_4)\right)\mu_4(ds_1\dots ds_4) + O(h^3). \qquad (13.32)\end{aligned}$$

Note that $\delta_k^0(t)$ and $\delta_k^1(t)$ coincide outside (t_k, t_{k+1}), $k = 0\dots, n-1$. In particular, they coincide at the nodes t_k, $k = 0,\dots,n$. Consider the set $\mathcal{R}_k$ of all points (s_1, s_2, s_3, s_4) in $[t_0, T]^4$ at which the products $\delta_k^0(s_1)\dots\delta_k^0(s_4)$ and $\delta_k^1(s_1)\dots\delta_k^1(s_4)$ are different (the Lebesgue measure of this set is $O(h)$). We will assume that f^4 is uniformly bounded while $\mu_4(\mathcal{R}_k)$ is $O(h)$ uniformly in k and n (this assumption does not exclude the possibility that the measure μ_4 is concentrated on manifolds of dimensions 1, 2, and 3 and also, e.g., at the points $(0,0,0,0)$ and (T,T,T,T)). Then the second integral in (13.35) is $O(h^3)$.

To prove (13.30) it remains to show that (under appropriate assumptions) the integral in (13.35) over the square $[t_0, T]^2$ is $O(h^3)$. We will assume that μ_2 is Lebesgue measure on $[t_0, T]^2$ plus Lebesgue measure concentrated on the diagonal of this square joining the points $(0,0)$ and (T,T), plus measure concentrated at finitely many points (t_k, t_k) that do not change with increasing n (e.g., the points $(0,0)$ and (T,T), or the midpoint of the above-mentioned diagonal in case each subsequent partition of $[0,T]$ is obtained from the previous one by dividing the intervals in two halves).

The expression $\delta_k^1(s_1)\delta_k^1(s_2) - \delta_k^0(s_1)\delta_k^0(s_2)$ vanishes outside the set P_k that is the difference of the squares $(t_k, T]^2$ and $[t_{k+1}, T]^2$. The set P_k does not contain points of the form (t_i, t_i). Therefore integration with respect to the point measures gives zero. Further, integration with respect to the measure concentrated on the diagonal leads to the integral

$$J_k = \int\limits_{t_k}^{t_{k+1}} f_k^2(s)\mathbf{E}\left(\delta_k^{1^2}(s) - \delta_k^{0^2}(s)\right) ds,$$

where $f_k^2(s)$ is a notation for $f^2(\overline{w}_k; s, s)$.

We have:

$$\mathbf{E}\left(\delta_k^{1^2}(s) - \delta_k^{0^2}(s)\right) = \begin{cases} s - t_k, & t_k < s < t_k + \frac{h}{2}, \\ s - t_{k+1}, & t_k + \frac{h}{2} \leq s < t_{k+1}. \end{cases}$$

Assume that $f_k^2(s)$ has a uniformly bounded derivative for $s \in (t_k, t_{k+1})$. Integrating by parts in the integral J_k gives

$$J_k = -\int\limits_{t_k}^{t_k+h/2} \frac{(s-t_k)^2}{2}\frac{df_k^2(s)}{ds}\, ds - \int\limits_{t_k+h/2}^{t_{k+1}} \frac{(s-t_{k+1})^2}{2}\frac{df_k^2(s)}{ds}\, ds.$$

This and our assumptions imply that integration with respect to the measure concentrated on the diagonal contributes $O(h^3)$. For integrating with respect to the two-dimensional Lebesgue measure we partition the set P_k into the square $(t_k, t_{k+1})^2$ and the two rectangles $(t_k, t_{k+1}) \times (t_{k+1}, T]$ and $(t_{k+1}, T] \times (t_k, t_{k+1})$. Integration over the square gives $O(h^3)$, since the area is h^2. The integral over, e.g., the first rectangle is equal to

$$\int\limits_{t_{k+1}}^{T} ds_2 \left[\int\limits_{t_k}^{t_k+h/2} f^2(\overline{w}_k; s_1, s_2)(s_1 - t_k)\, ds_1 + \int\limits_{t_k+h/2}^{t_{k+1}} f^2(\overline{w}_k; s_1, s_2)(s_1 - t_{k+1})\, ds_1\right],$$

so that (recall that $\delta_k^1(s_2) = \delta_k^0(s_2) = \Delta w_k$ for $t_{k+1} < s_2 \leq T$)

$$\mathbf{E}\left(\delta_k^1(s_1)\delta_k^1(s_2) - \delta_k^0(s_1)\delta_k^0(s_2)\right) = \mathbf{E}\left(\delta_k^1(s_1) - \delta_k^0(s_1)\right)\delta_k^0(s_2)$$

$$= \begin{cases} \mathbf{E}\left(w(s_1) - w(t_k)\right)\left(w(t_{k+1}) - w(t_k)\right), & t_k < s_1 < t_k + \frac{h}{2}, \quad t_{k+1} < s_2 \leq T, \\ \mathbf{E}\left(w(s_1) - w(t_{k+1})\right)\left(w(t_{k+1}) - w(t_k)\right), & t_k + \frac{h}{2} \leq s_1 < t_{k+1}, \quad t_{k+1} < s_2 \leq T, \end{cases}$$

$$= \begin{cases} s_1 - t_k, & t_k < s_1 < t_k + \frac{h}{2}, \quad t_{k+1} < s_2 \leq T, \\ s_1 - t_{k+1}, & t_k + \frac{h}{2} \leq s_1 < t_{k+1}, \quad t_{k+1} < s_2 \leq T. \end{cases}$$

The assumption that the derivative $\partial f^2(\overline{w}_k; s_1, s_2)/\partial s_1$ exists and is bounded allows us to state as conclusion that this integral contributes $O(h^3)$. The reasoning is similar as done for J_k. A similar assumption and derivation can be given for the integral over the second rectangle. So, (13.30) has been completely substantiated.

EXAMPLE 13.1. Let $y_t(x(\cdot))$ be a solution of the integral equation

$$y_t(x(\cdot)) = x(t) + \int\limits_0^t a(y_s(x(\cdot)))\, ds, \tag{13.33}$$

where a is a function with bounded continuous derivatives up to order six, inclusively.

Upon replacing x by a Wiener process w, (13.33) becomes a stochastic equation for a diffusion process with constant diffusion coefficient. All conditions mentioned in this Section are satisfied for, e.g., functionals of the form $F(x) = \varphi(y_T(x(\cdot)))$ or $F(x) = \varphi\left(y_T(x(\cdot)), \int_0^T f(s, y_s(x(\cdot)))\,ds\right)$ with sufficiently smooth functions f and φ. Thus, the Monte-Carlo methods based on the approximations of second order of accuracy treated in this Section can be used to compute the expectations $\mathbf{E}F(x)$ of such functionals.

REMARK 13.1. Note that the piecewise-linear approximation $w_h(t)$ of a Wiener process, defined as

$$w_h(t) = w(t_k)\frac{t_{k+1}-t}{h} + w(t_{k+1})\frac{t-t_k}{h}, \qquad t_k \le t \le t_{k+1},$$

and differing from the piecewise-linear approximation $w^h(t)$ used in the trapezium method, gives an error of order $O(h)$. To confirm the above said it is simplest to compute

$$\mathbf{E}\int_0^1 w^2(t)\,dt = \frac{1}{2}, \qquad \mathbf{E}\int_0^1 w^{h^2}(t)\,dt = \frac{1}{2}, \qquad \mathbf{E}\int_0^1 w_h^2(t)\,dt = \frac{1}{2} - \frac{h}{6}.$$

14. Methods of order of accuracy four for computing Wiener integrals of functionals of exponential type

14.1. Introduction. This Section is devoted to the computation of *Wiener integrals* (13.1) of often encountered functionals of exponential type

$$V(x(\cdot)) = \exp\left(\int_0^T a(t, x(t))\,dt\right), \qquad x(t) \in \mathbb{R}^n. \tag{14.1}$$

The functional (14.1) is a particular case of the functional (13.2), and therefore the results of the previous Section can be applied here too. In §13.4 we have noted that there is no Runge–Kutta method of order of accuracy exceeding two for integrating the system (13.3). However, thanks to the special form of the functional (14.1), the computation of the integrals (13.1), (14.1) can be done by using another system, for which we can successfully develop a method of order of accuracy four. This system has the form

$$\begin{aligned} dX^1(s) &= dw^1(s), \qquad X^1(t) = x^1, \\ &\cdots\cdots\cdots \\ dX^n(s) &= dw^n(s), \qquad X^n(t) = x^n, \\ dY(s) &= Y(s)a(s, X^1(s), \ldots, X^n(s))\,ds, \qquad Y(t) = y, \end{aligned} \tag{14.2}$$

where $0 \le t \le s$.

We will denote the solution of this system by $X_{t,x}(s)$, $Y_{t,x,y}(s)$ or, if this does not give rise to confusion, by $X(s)$, $Y(s)$.

It can be readily seen that the Wiener integral of the functional (14.1) is equal to

$$I = \mathbf{E}Y(T),$$

where $Y(s)$ is the corresponding coordinate of the solution of the system (14.2) with initial conditions $t = 0$, $x^1 = \cdots = x^n = 0$, $y = 1$.

Thus, the search for the Winer integral (13.1), (14.1) leads to the problem of numerically integrating the system (14.2). Let $\overline{Y}(s)$ be the coordinate of an approximate solution. For our purposes the approximate solution becomes better with diminishing difference of mathematical expectations $\mathbf{E}Y(t) - \mathbf{E}\overline{Y}(t)$. Despite the facts that the $X^1(s), \dots, X^n(s)$ in (14.2) can be found exactly and that the equation for $Y(s)$ does not have stochastic components, as in §13 we need special proofs for using methods that are well known in the deterministic case. However, in our case the example given in §13.4 is not completely convincing. In fact, in that Section the difference involved was that between $\mathbf{E}Z^2(h)$ and $\mathbf{E}\overline{Z}^2(h)$, while at the same time $\mathbf{E}Z(h) = \mathbf{E}\overline{Z}(h)$. Here, however, it is sufficient to construct $\overline{Y}$ such that $\mathbf{E}\overline{Y}$ differs but little from $\mathbf{E}Y$. Therefore we give a new example, confirming the need of a separate substantiation.

To this end we consider the Cauchy problem for the one-dimensional equation

$$dx = \frac{1 + \left(1 + \int_0^t w(s)\,ds\right)^2}{1 + (1 + x)^2} w(t)\,dt, \qquad x(0) = 0.$$

It has solution $x(t) = \int_0^t w(s)\,ds$. Clearly $\mathbf{E}x(h) = 0$. To approximately find $x(h)$ we use a Runge–Kutta method of third order of accuracy (see [2, pp. 300–303]):

$$f(t, x) = \frac{1 + \left(1 + \int_0^t w(s)\,ds\right)^2}{1 + (1 + x)^2} w(t), \qquad t_0 = 0, \qquad x_0 = 0,$$

$$k_1 = hf(t_0, x_0) = 0,$$

$$k_2 = hf(t_0 + \alpha_2 h, x_0 + \beta_{21} k_1) = h \frac{1 + \left(1 + \int_0^{\alpha_2 h} w(s)\,ds\right)^2}{2} w(\alpha_2 h),$$

$$k_3 = hf(t_0 + \alpha_3 h, x_0 + \beta_{31} k_1 + \beta_{32} k_2) = h \frac{1 + \left(1 + \int_0^{\alpha_3 h} w(s)\,ds\right)^2}{1 + (1 + \beta_{33} k_2)^2} w(\alpha_3 h),$$

$$\overline{x}(h) = x_0 + p_{31} k_1 + p_{32} k_2 + p_{33} k_3,$$

$$\mathbf{E}\overline{x}(h) = x_0 + p_{31} \mathbf{E}k_1 + p_{32} \mathbf{E}k_2 + p_{33} \mathbf{E}k_3 = p_{32} \mathbf{E}k_2 + p_{33} \mathbf{E}k_3. \tag{14.3}$$

We can immediately compute that

$$\mathbf{E}k_2 = \frac{1}{2}\alpha_2^2 h^3 + O(h^4),$$

$$\mathbf{E}k_3 = \frac{1}{2}\alpha_3^2 h^3 - \beta_{32} \min(\alpha_1, \alpha_2) h^3 + O(h^4). \tag{14.4}$$

The parameters of the method satisfy the system of equations

$$\alpha_2 = \beta_{21}, \quad \alpha_3 = \beta_{31} + \beta_{32}, \quad p_{31} + p_{32} + p_{33} = 1, \quad p_{32}\alpha_2 + p_{33}\alpha_3 = \frac{1}{2},$$
$$p_{32}\alpha_2^2 + p_{33}\alpha_3^2 = \frac{1}{3}, \quad p_{33}\beta_{32}\alpha_2 = \frac{1}{6}.$$

We choose the following parameter values:

$$\alpha_2 = \beta_2 = 1, \quad \alpha_3 = \frac{1}{2}, \quad \beta_{31} = \beta_{32} = \frac{1}{4}, \quad p_{31} = p_{32} = \frac{1}{6}, \quad p_{33} = \frac{2}{3}.$$

Substituting these values into (14.4), and then in (14.3) we obtain $\mathbf{E}\overline{x}(h) = h^3/12 + O(h^4)$, which differs from $\mathbf{E}x(h) = 0$ by $O(h^3)$ instead of by the 'obviously' expected $O(h^4)$. It is clear that the mismatch with the deterministic case occurs because of the nonsmoothness of $w(t)$.

In the next Subsection we substantiate a Runge–Kutta method of fourth order of accuracy for integrating the system (14.2) In §14.3 we will consider the problem of reducing the probability error of the Monte-Carlo method for computing the Wiener integral of the functional (14.2). As is well known, this error is proportional to $(\mathbf{V}Y(T)/N)^{1/2}$. As in §12, to reduce it we construct another variable, $\widetilde{Y}(T)$, such that $\mathbf{E}\widetilde{Y}(T) = \mathbf{E}Y(T)$ while at the same time $\mathbf{V}\widetilde{Y}(T) < \mathbf{V}Y(T)$. We can construct the system of stochastic differential equations producing the required random variable $\widetilde{Y}(T)$ similarly as in §12. In §14.4 we give the results of some numerical experiments.

14.2. A fourth-order Runge–Kutta method for integrating the system (14.2). We will denote an approximate solution of the system (14.2) on the interval $t \le s \le T$ by $\overline{X}_{t,x}(s) = (\overline{X}^1_{t,x}(s), \dots, \overline{X}^n_{t,x}(s))$, $\overline{Y}_{t,x,y}(s)$ (or by $\overline{X}(s)$, $\overline{Y}(s)$). We look for it at the points $s_k = t + kh$, $k = 0, \dots, m$; $h = (T - t)/m$.

Consider the following approximation:

$$\overline{X}(t) = x, \qquad \overline{X}(s_{k-1/2}) = \overline{X}(s_{k-1}) + \frac{1}{\sqrt{2}}\xi_{k-1/2}h^{1/2},$$
$$\overline{X}(s_k) = \overline{X}(s_{k-1/2}) + \frac{1}{\sqrt{2}}\xi_k h^{1/2},$$
$$\overline{Y}(t) = y,$$
$$\overline{Y}(s_k) = \overline{Y}(s_{k-1}) + \frac{1}{6}\left[k_1(h) + 2k_2(h) + 2k_3(h) + k_4(h)\right], \tag{14.5}$$

where

$$k_1(h) = ha\left(s_{k-1}, \overline{X}(s_{k-1})\right)\overline{Y}(s_{k-1}),$$
$$k_2(h) = ha\left(s_{k-1/2}, \overline{X}(s_{k-1/2})\right)\left[\overline{Y}(s_{k-1}) + \frac{k_1(h)}{2}\right],$$
$$k_3(h) = ha\left(s_{k-1/2}, \overline{X}(s_{k-1/2})\right)\left[\overline{Y}(s_{k-1}) + \frac{k_2(h)}{2}\right],$$
$$k_4(h) = ha\left(s_k, \overline{X}(s_k)\right)\left[\overline{Y}(s_{k-1}) + k_3(h)\right], \tag{14.6}$$

where the ξ_j (j a half-integer) are mutually independent n-vectors consisting of independent $N(0,1)$-distributed components.

Introduce the functions

$$\varphi(t,x) = \mathbf{E}\exp\left(\int_t^T a(s, X_{t,x}(s))\,ds\right),$$

$$u(t,x,y) = \mathbf{E}Y_{t,x,y}(T) = y\mathbf{E}\exp\left(\int_t^T a(s, X_{t,x}(s))\,ds\right) = y\varphi(t,x).$$

The main purpose of this Subsection is to prove the relation

$$\mathbf{E}\left(\left(\overline{Y}_{t,x,y}(t+h) - Y_{t,x,y}(t+h)\right)\varphi(t+h, X_{t,x}(t+h))\right) = O(h^5). \tag{14.7}$$

This relation implies that the method (14.5), (14.6) has fourth order of accuracy:

$$\mathbf{E}\left(\overline{Y}_{t,x,y}(T) - Y_{t,x,y}(T)\right) = O(h^4). \tag{14.8}$$

To prove this we can proceed as in Theorem 9.1. Here, the role of the one-step error $\mathbf{E}u(t_{i+1}, X_{t_i,\overline{X}_i}(t_i+h)) - \mathbf{E}u(t_{i+1}, \overline{X}_{t_i,\overline{X}_i}(t_i+h))$ in (9.12) is played by the lefthand side of (14.7). Recall that instead of X ($\overline{X}$) in §9, here we have the pair (X,Y) ($(\overline{X},\overline{Y})$), that a component $\overline{X}$ of this pair can be modeled exactly, and that Y ($\overline{Y}$) is homogeneous with respect to the initial component y.

Note that below we will apply Lemma 13.1 to the function $u(t,x,y) = y\varphi(t,x)$. This is possible if, e.g., a is nonpositive and sufficiently smooth.

We begin with the proof of (14.7). To this end we turn to the system (14.2). Its operator L applied to the function $\psi(t,x,y) = y\varphi(t,x)$ gives

$$L[y\varphi] = y(L_0\varphi + a\varphi), \tag{14.9}$$

where

$$L_0 = \frac{\partial}{\partial t} + \Delta = \frac{\partial}{\partial t} + \frac{1}{2}\sum_{i=1}^n \frac{\partial^2}{\partial x^{i2}}.$$

By (13.14),

$$\mathbf{E}\left[Y_{t,x,y}(t+h)\varphi(t+h, X_{t,x}(t+h))\right] = y\varphi(x) + hL[y\varphi] + \frac{h^2}{2}L^2[y\varphi] + \frac{h^3}{6}L^3[y\varphi] + \frac{h^4}{24}L^4[y\varphi] + O(h^5). \tag{14.10}$$

In this formula $L[y\varphi]$ can be computed by (14.9) and $L^2[y\varphi]$, $L^3[y\varphi]$, $L^4[y\varphi]$ by the formulas (where we do not write the argument (t,x)):

$$
\begin{aligned}
L^2[y\varphi] &= y\left(L_0^2\varphi + L_0[a\varphi] + aL_0\varphi + a^2\varphi\right),\\
L^3[y\varphi] &= y\left(L_0^3\varphi + L_0^2[a\varphi] + L_0\left[aL_0\varphi\right] + L_0\left[a^2\varphi\right] + aL_0^2\varphi\right.\\
&\quad \left. + aL_0\left[f\varphi\right] + a^2L_0\varphi + a^3\varphi\right),
\end{aligned}
$$

$$
\begin{aligned}
L^4[y\varphi] = y&\left(L_0^4\varphi + L_0^4[a\varphi] + L_0^2\left[aL_0\varphi\right] + L_0^2\left[a^2\varphi\right] + L_0\left[aL_0^2\varphi\right]\right.\\
&+ L_0\left[aL_0\left[a\varphi\right]\right] + L_0\left[a^2L_0\varphi\right] + L_0\left[a^3\varphi\right] + aL_0^3\varphi\\
&\left. + aL_0^2\left[a\varphi\right] + aL_0\left[aL_0\varphi\right] + aL_0\left[a^2\varphi\right] + a^2L_0^2\varphi + a^2L_0\left[a\varphi\right] + a^3L_0\varphi + a^4\varphi\right).
\end{aligned}
$$

Using (14.5), (14.6) we can immediately compute

$$
\begin{aligned}
&\mathbf{E}\left\{\overline{Y}_{t,x,y}(t+h)\varphi(t+h, X_{t,x}(t+h))\right\} = y\mathbf{E}\varphi(t+h, X_{t,x}(t+h))\\
&+ y\frac{h}{6}\left[a(t,x)\mathbf{E}\varphi(t+h, X_{t,x}(t+h))\right.\\
&+ 4\mathbf{E}a\left(t+\frac{h}{2}, X_{t,x}\left(t+\frac{h}{2}\right)\right)\varphi(t+h, X_{t,x}(t+h))\\
&\left. + \mathbf{E}a(t+h, X_{t,x}(t+h))\varphi(t+h, X_{t,x}(t+h))\right]\\
&+ y\frac{h^2}{6}\left[a(t,x)\mathbf{E}a\left(t+\frac{h}{2}, X_{t,x}\left(t+\frac{h}{2}\right)\right)\varphi(t+h, X_{t,x}(t+h))\right.\\
&+ \mathbf{E}a^2\left(t+\frac{h}{2}, X_{t,x}\left(t+\frac{h}{2}\right)\right)\varphi(t+h, X_{t,x}(t+h))\\
&\left. + \mathbf{E}a\left(t+\frac{h}{2}, X_{t,x}\left(t+\frac{h}{2}\right)\right)a(t+h, X_{t,x}(t+h))\varphi(t+h, X_{t,x}(t+h))\right]\\
&+ y\frac{h^3}{12}\left[a(t,x)\mathbf{E}a^2\left(t+\frac{h}{2}, X_{t,x}\left(t+\frac{h}{2}\right)\right)\varphi(t+h, X_{t,x}(t+h))\right.\\
&\left. + \mathbf{E}a^2\left(t+\frac{h}{2}, X_{t,x}\left(t+\frac{h}{2}\right)\right)a(t+h, X_{t,x}(t+h))\varphi(t+h, X_{t,x}(t+h))\right]\\
&+ y\frac{h^4}{24}a(t,x)\mathbf{E}a^2\left(t+\frac{h}{2}, X_{t,x}\left(t+\frac{h}{2}\right)\right)a(t+h, X_{t,x}(t+h))\varphi(t+h, X_{t,x}(t+h)).
\end{aligned}
\tag{14.11}
$$

By Lemma 13.1 (formula (13.14)),

$$
\mathbf{E}\varphi(t+h, X_{t,x}(t+h)) = \varphi + hL_0\varphi + \frac{h^2}{2}L_0^2\varphi + \frac{h^3}{6}L_0^3\varphi + \frac{h^4}{24}L_0^4\varphi + O(h^5).
\tag{14.12}
$$

It suffices to compute the expression $\mathbf{E}a\left(t+\frac{h}{2}, X_{t,x}\left(t+\frac{h}{2}\right)\right)\varphi(t+h, X_{t,x}(t+h))$ up to $O(h^4)$:

$$\begin{aligned}
&\mathbf{E}a\left(t+\frac{h}{2}, X_{t,x}\left(t+\frac{h}{2}\right)\right)\varphi(t+h, X_{t,x}(t+h)) \\
&\quad = \mathbf{E}\mathbf{E}\left\{a\left(t+\frac{h}{2}, X_{t,x}\left(t+\frac{h}{2}\right)\right)\varphi(t+h, X_{t,x}(t+h)) \mid \mathcal{F}_{t+h/2}\right\} \\
&\quad = \mathbf{E}\left(a\left(t+\frac{h}{2}, X_{t,x}\left(t+\frac{h}{2}\right)\right)\mathbf{E}\left\{\varphi(t+h, X_{t,x}(t+h)) \mid \mathcal{F}_{t+h/2}\right\}\right).
\end{aligned} \tag{14.13}$$

By (13.11), the conditional mathematical expectation at the righthand side of (14.13) can be written as

$$\begin{aligned}
\mathbf{E}\left\{\varphi(t+h, X_{t,x}(t+h)) \mid \mathcal{F}_{t+h/2}\right\} &= \varphi\left(t+\frac{h}{2}, X_{t,x}\left(t+\frac{h}{2}\right)\right) \\
&\quad + \frac{h}{2}L_0\varphi\left(t+\frac{h}{2}, X_{t,x}\left(t+\frac{h}{2}\right)\right) \\
&\quad + \left(\frac{h}{2}\right)^2 \cdot \frac{1}{2}L_0^2\varphi\left(t+\frac{h}{2}, X_{t,x}\left(t+\frac{h}{2}\right)\right) \\
&\quad + \left(\frac{h}{2}\right)^3 \cdot \frac{1}{6}L_0^3\varphi\left(t+\frac{h}{2}, X_{t,x}\left(t+\frac{h}{2}\right)\right) + O(h^4).
\end{aligned}$$

Substituting this into (14.13) gives

$$\begin{aligned}
&\mathbf{E}a\left(t+\frac{h}{2}, X_{t,x}\left(t+\frac{h}{2}\right)\right)\varphi(t+h, X_{t,x}(t+h)) \\
&\quad = \mathbf{E}a\left(t+\frac{h}{2}, X_{t,x}\left(t+\frac{h}{2}\right)\right)\varphi\left(t+\frac{h}{2}, X_{t,x}\left(t+\frac{h}{2}\right)\right) \\
&\quad + \frac{h}{2}\mathbf{E}a\left(t+\frac{h}{2}, X_{t,x}\left(t+\frac{h}{2}\right)\right)L_0\varphi\left(t+\frac{h}{2}, X_{t,x}\left(t+\frac{h}{2}\right)\right) \\
&\quad + \frac{h^2}{8}\mathbf{E}a\left(t+\frac{h}{2}, X_{t,x}\left(t+\frac{h}{2}\right)\right)L_0^2\varphi\left(t+\frac{h}{2}, X_{t,x}\left(t+\frac{h}{2}\right)\right) \\
&\quad + \frac{h^3}{48}\mathbf{E}a\left(t+\frac{h}{2}, X_{t,x}\left(t+\frac{h}{2}\right)\right)L_0^3\varphi\left(t+\frac{h}{2}, X_{t,x}\left(t+\frac{h}{2}\right)\right) + O(h^4).
\end{aligned} \tag{14.14}$$

Each mathematical expectation at the righthand side of (14.14) can, by Lemma 13.1, be expanded with respect to powers of h. Here, the first mathematical expectation has to be expanded up to $O(h^4)$, the second up to $O(h^3)$, the third up to $O(h^2)$, and the fourth up to $O(h)$:

$$\begin{aligned}
&\mathbf{E}a\left(t+\frac{h}{2}, X_{t,x}\left(t+\frac{h}{2}\right)\right)\varphi\left(t+\frac{h}{2}, X_{t,x}\left(t+\frac{h}{2}\right)\right)\\
&\quad = a\varphi + \frac{h}{2}L_0\left[a\varphi\right] + \frac{h^2}{8}L_0^2\left[a\varphi\right] + \frac{h^3}{48}L_0^3\left[a\varphi\right] + O(h^4),\\
&\mathbf{E}a\left(t+\frac{h}{2}, X_{t,x}\left(t+\frac{h}{2}\right)\right)L_0\varphi\left(t+\frac{h}{2}, X_{t,x}\left(t+\frac{h}{2}\right)\right)\\
&\quad = aL_0\varphi + \frac{h}{2}L_0\left[aL_0\varphi\right] + \frac{h^2}{8}L_0^2\left[aL_0\varphi\right] + O(h^3),\\
&\mathbf{E}a\left(t+\frac{h}{2}, X_{t,x}\left(t+\frac{h}{2}\right)\right)L_0^2\varphi\left(t+\frac{h}{2}, X_{t,x}\left(t+\frac{h}{2}\right)\right)\\
&\quad = aL_0^2\varphi + \frac{h}{2}L_0\left[aL_0^2\varphi\right] + O(h^2),\\
&\mathbf{E}a\left(t+\frac{h}{2}, X_{t,x}\left(t+\frac{h}{2}\right)\right)L_0^3\varphi\left(t+\frac{h}{2}, X_{t,x}\left(t+\frac{h}{2}\right)\right) = aL_0^3\varphi + O(h^3).
\end{aligned} \tag{14.15}$$

Finally, substituting (14.15) into (14.14) we obtain

$$\begin{aligned}
&\mathbf{E}a\left(t+\frac{h}{2}, X_{t,x}\left(t+\frac{h}{2}\right)\right)\varphi\left(t+h, X_{t,x}(t+h)\right) = a\varphi + \frac{h}{2}\left(L_0\left[a\varphi\right] + aL_0^2\varphi\right)\\
&\quad + \frac{h^2}{8}\left(L_0^2\left[a\varphi\right] + 2L_0\left[aL_0\varphi\right] + aL_0\varphi\right)\\
&\quad + \frac{h^3}{48}\left(L_0^3\left[a\varphi\right] + 3L_0^2\left[aL_0\varphi\right] + 3L_0\left[aL_0^2\varphi\right] + aL_0^3\varphi\right) + O(h^4).
\end{aligned} \tag{14.16}$$

Similar reasonings allow us to obtain the following formulas for the mathematical expectation of the righthand sides in (14.11) with the required accuracy:

$$\begin{aligned}
&\mathbf{E}a(t+h, X_{t,x}(t+h))\varphi(t+h, X_{t,x}(t+h)) = a\varphi + hL_0\,[a\varphi] \\
&\quad + \frac{h^2}{2}L_0^2\,[a\varphi] + \frac{h^3}{6}L_0^3\,[a\varphi] + O(h^4), \\
&\mathbf{E}a^2\left(t+\frac{h}{2}, X_{t,x}\left(t+\frac{h}{2}\right)\right)\varphi(t+h, X_{t,x}(t+h)) = a^2\varphi + \frac{h}{2}\left(L_0\left[a^2\varphi\right] + a^2L_0\varphi\right) \\
&\quad + \frac{h^2}{8}\left(L_0^2\left[a^2\varphi\right] + 2L_0\left[a^2L_0\varphi\right] + a^2L_0^2\varphi\right) + O(h^3), \\
&\mathbf{E}a\left(t+\frac{h}{2}, X_{t,x}\left(t+\frac{h}{2}\right)\right) a(t+h, X_{t,x}(t+h))\varphi(t+h, X_{t,x}(t+h)) \\
&\quad = a^2\varphi + \frac{h}{2}\left(L_0\left[a^2\varphi\right] + aL_0\,[a\varphi]\right) \\
&\quad + \frac{h^2}{8}\left(L_0^2\left[a^2\varphi\right] + 2L_0\,[aL_0\,[a\varphi]] + aL_0^2\,[a\varphi]\right) + O(h^3), \\
&\mathbf{E}a^2\left(t+\frac{h}{2}, X_{t,x}\left(t+\frac{h}{2}\right)\right) a(t+h, X_{t,x}(t+h))\varphi(t+h, X_{t,x}(t+h)) \\
&\quad = a^3\varphi + \frac{h}{2}\left(L_0\left[a^3\varphi\right] + a^2L_0\,[a\varphi]\right) + O(h^2).
\end{aligned} \tag{14.17}$$

Substituting (14.12), (14.16), and (14.17) into (14.11) we find the expansion of $\mathbf{E}\left\{\overline{Y}_{t,x,y}(t+h)\varphi(t+h, X_{t,x}(t+h))\right\}$ with respect to powers of h up to $O(h^5)$.

We can immediately convince ourselves that this expansion coincides with that for $\mathbf{E}\left\{Y_{t,x,y}(t+h)\varphi(t+h, X_{t,x}(t+h))\right\}$. Thus, (14.17) has been proved and, so, the method (14.5), (14.6) has fourth order of accuracy from the point of view of computing $\mathbf{E}Y(T)$.

REMARK 14.1. It is not difficult to see that the method (14.5), (14.6) is a Runge–Kutta method of fourth order of accuracy for integrating the last equation in the system (14.2) if we consider $a(s, w_1(s), \ldots, w_n(s))$ to be a known function of s. Since this function is random and nonsmooth, as already mentioned in §14.1 we need a separate proof. We have given this proof here. We can also obtain Runge–Kutta methods of second or third order of accuracy. Note that the method (14.5), (14.6) requires at each step only a double modeling of a normally-distributed random vector and a double computation of the function a. The method of third order of accuracy requires the same amount of computations as does the method (14.5), (14.6), while the method of second order of accuracy, which is already quite inferior as regards accuracy, requires only a somewhat less amount of computations. Therefore the Runge–Kutta methods of third and second order of accuracy are in this case not of interest.

14.3. Reducing variances. In relation with the computation of a Wiener integral of a functional (14.1) there arises the problem of computing $\mathbf{E}Y(T)$ via the system (14.2). By approximately integrating the system (14.2) by some method, we replace the computation of $\mathbf{E}Y(T)$ by that of $\mathbf{E}\overline{Y}(T)$, where $\overline{Y}$ is an approximate solution. If $\mathbf{E}\overline{Y}(T)$ is to be computed by the Monte-Carlo method, then by performing N independent trials (for this we have to find approximately N trajectories and to obtain values $\overline{Y}^{(m)}(T)$, $m = 1, \dots, N$) we find

$$\mathbf{E}\overline{Y}(T) \doteq \frac{1}{N} \sum_{m=1}^{N} \overline{Y}^{(m)}(T). \tag{14.18}$$

The error in the relation (14.18) (the *error of the Monte-Carlo method*) can be estimated by $\left(\mathbf{V}\overline{Y}(T)/N\right)^{1/2}$. Of course, we assume that all mathematical expectations and variances under consideration exist.

We will show that for the method (14.5), (14.6),

$$\mathbf{V}\overline{Y}(T) = \mathbf{V}Y(T) + O(h). \tag{14.19}$$

For this we have to prove the following property of one-step approximations (see the proof of Theorem 9.1):

$$\mathbf{E}\psi\left(\overline{X}_{t,x}(t+h), \overline{Y}_{t,x,y}(t+h)\right) = \mathbf{E}\psi\left(X_{t,x}(t+h), Y_{t,x,y}(t+h)\right) + O(h^2). \tag{14.20}$$

We have:

$$\mathbf{E}\psi\left(X_{t,x}(t+h), Y_{t,x,y}(t+h)\right) = \psi + hL\psi + O(h^2),$$

$$\begin{aligned}
\mathbf{E}\psi & \left(\overline{X}_{t,x}(t+h), \overline{Y}_{t,x,y}(t+h)\right) = \mathbf{E}\psi\Bigg(X_{t,x}(t+h), y + \frac{1}{6}\Bigg[hay \\
&+ 2ha\left(t+\frac{h}{2}, X_{t,x}\left(t+\frac{h}{2}\right)\right)\left(y + \frac{1}{2}hay\right) \\
&+ 2ha\left(t+\frac{h}{2}, X_{t,x}\left(t+\frac{h}{2}\right)\right)\left(y + \frac{1}{2}ha\left(t+\frac{h}{2}, X_{t,x}\left(t+\frac{h}{2}\right)\right)\left(y+\frac{1}{2}hay\right)\right) \\
&+ ha(t+h, X_{t,x}(t+h))\left(y + ha\left(t+\frac{h}{2}, X_{t,x}\left(t+\frac{h}{2}\right)\right)\right. \\
&\times \left.\left.\left.\left(y + \frac{1}{2}ha\left(t+\frac{h}{2}, X_{t,x}\left(t+\frac{h}{2}\right)\right)\left(y+\frac{1}{2}hay\right)\right)\right)\right]\Bigg) \\
&= \mathbf{E}\psi\left(X_{t,x}(t+h), y + h\left(\frac{1}{6}ay + \frac{2}{3}a\left(t+\frac{h}{2}, X_{t,x}\left(t+\frac{h}{2}\right)\right)y\right.\right. \\
&\left.\left. + \frac{1}{6}a(t+h, X_{t,x}(t+h))y\right)\right) + O(h^2) \\
&= \mathbf{E}\left(\psi(x,y) + \sum_{i=1}^{n} \frac{\partial \psi}{\partial x^i}\Delta w^i(h) + \frac{1}{2}\sum_{i=1}^{n}\frac{\partial^2\psi}{\partial x^{i2}}\left(\Delta w^i(h)\right)^2 + \psi'_y hay\right) + O(h^2) \\
&= \psi + hL\psi + O(h^2).
\end{aligned}$$

Thus, (14.20) has been proved. Note that this relation, and hence also (14.19), holds also for the Euler method.

If $\mathbf{V}Y(T)$ is large, then by (14.19) $\mathbf{V}\overline{Y}(T)$ is also large and to achieve a satisfactory accuracy in (14.18) we have to compute a very large amount of trajectories. If instead of $Y(T)$ we succeed in constructing another variable, $\widetilde{Y}(T)$ such that $\mathbf{E}\widetilde{Y}(T) = \mathbf{E}Y(T)$ and $\mathbf{V}\widetilde{Y}(T) < \mathbf{V}Y(T)$, then modeling $\widetilde{Y}(T)$ instead of $Y(T)$ would allow us to obtain more refined results with the same expense of computations. Below we largely repeat the construction of §12.

Consider the system (14.2).

Introduce the function

$$u(t, x^1, \ldots, x^n)y = \mathbf{E}Y_{t,x^1,\ldots,x^n}(T) \tag{14.21}$$

(it is clearly linear and homogeneous with respect to y, which we have reflected immediately in the notation). By §14.1, the value of this function at $t = 0$, $x^1 = \cdots = x^n = 0$, $y = 1$ is the value of the Wiener integral of the functional (14.1) looked for. The function $u(t, x^1, \ldots, x^n)$ satisfies the equation

$$\frac{\partial u}{\partial t} + \frac{1}{2}\Delta u + au = 0 \tag{14.22}$$

and the initial condition

$$u(T, x^1, \ldots, x^n) = 1. \tag{14.23}$$

Instead of (14.2) we consider the system

$$\begin{aligned}
dX^1(s) &= -\mu^1(s, X^1, \ldots, X^n)\,ds + dw^1(s), \qquad X^1(t) = x^1,\\
&\cdots\cdots\cdots\cdots\cdots\cdots\cdots\cdots\\
dX^n(s) &= -\mu^n(s, X^1, \ldots, X^n)\,ds + dw^n(s), \qquad X^n(t) = x^n,\\
dY(s) &= Ya(s, X^1, \ldots, X^n)\,ds + \mu^1 Y\,dw^1 + \cdots + \mu^n Y\,dw^n, \qquad Y(t) = y.
\end{aligned} \tag{14.24}$$

If we introduce u by (14.21) with Y found from (14.24), then it is easy to verify that this function again satisfies (14.22), (14.23). Therefore (naturally, we assume that the solution of the problem (14.22), (14.23) exists and is unique) the value of $\mathbf{E}Y(T)$, computed along the system (14.24), does not depend on the choice of the functions $\mu^1, \ldots, \mu^n$ and so coincides with its value at $\mu^1 \equiv 0, \ldots, \mu^n \equiv 0$, i.e. with the value looked for.

At the same time $\mathbf{V}Y$ does depend on $\mu^1, \ldots, \mu^n$. In relation with this it is natural to regard $\mu^1, \ldots, \mu^n$ as controls and to choose them from the condition that the variance

$$\mathbf{V}Y(T) = \mathbf{E}Y^2(T) - (\mathbf{E}Y(T))^2$$

be minimal. Since $\mathbf{E}Y(T)$ is independent of $\mu^1, \ldots, \mu^n$, this choice reduces to solving the following problem from optimal control theory: it is required to choose controls $\mu^1, \ldots, \mu^n$ giving a minimum of the functional

$$I(\mu^1, \ldots, \mu^n) = \mathbf{E}Y^2(T) \tag{14.25}$$

along the system (14.24).

Introduce the variable $Z = Y^2$. By Itô's formula,

$$dZ = 2aZ\,ds + \sum_{i=1}^{n} \mu^{i^2} Z\,ds + 2\left(\sum_{i=1}^{n} \mu^i\,dw^i\right) Z.$$

As a result the problem of minimising the functional (14.25) is equivalent to that of minimising the functional

$$I(\mu^1, \dots, \mu^n) = \mathbf{E}Z(T)$$

along the solutions of the system

$$dX^1(s) = -\mu^1(s, X^1, \dots, X^n)\,ds + dw^1(s), \qquad X^1(t) = x^1,$$

$$\dots\dots\dots\dots\dots\dots\dots\dots\dots\dots\dots$$

$$dX^n(s) = -\mu^n(s, X^1, \dots, X^n)\,ds + dw^n(s), \qquad X^n(t) = x^n,$$

$$dZ(s) = \left(2a + \sum_{i=1}^{n} \mu^{i^2}\right) Z\,ds + 2Z\sum_{i=1}^{n} \mu^i\,dw^i, \qquad Z(t) = z. \tag{14.26}$$

Introduce the following function (it is clearly linear and homogeneous with respect to z):

$$v(t, x^1, \dots, x^n)z = \min_{\mu} \mathbf{E}Z_{t,x,z}(T).$$

THEOREM 14.1. *We have* $v = u^2$. *For*

$$\mu^i = -\frac{1}{2v}\frac{\partial v}{\partial x^i} = -\frac{\partial \ln u}{\partial x^i} \tag{14.27}$$

we have (*along the system* (14.24)) $\mathbf{E}Y_{t,x^1,\dots,x^n}(T) = 0$, *i.e. the variable* $Y(T)$, *computed according to* (14.24) *taking into account* (14.27), *is deterministic.*

PROOF. The function v satisfies the *Bellman equation* [9], [21], [40]

$$\min_{\mu}\left(\frac{\partial v}{\partial t} + \sum_{i=1}^{n} \mu^i \frac{\partial v}{\partial x^i} + \frac{1}{2}\sum_{i=1}^{n} \frac{\partial^2 v}{\partial x^{i^2}} + 2av + \sum_{i=1}^{n} \mu^{i^2} v\right) = 0 \tag{14.28}$$

and the condition $v(T, x^1, \dots, x^n) = 1$.

Formula (14.28) implies the first part of the identity (14.27) as well as the equation

$$\frac{\partial v}{\partial t} + \frac{1}{2}\sum_{i=1}^{n} \frac{\partial^2 v}{\partial x^{i^2}} + 2av - \frac{1}{4v}\sum_{i=1}^{n}\left(\frac{\partial v}{\partial x^i}\right)^2 = 0. \tag{14.29}$$

Changing variables $v = p^2$ in this equation leads to (14.22) with condition (14.23) relative to p. Thus, for the optimal μ^i the function v is in fact equal to u^2. This implies the second part of the identity (14.27), as well as $\mathbf{E}Y^2 = \mathbf{E}Z = v = u^2 = (\mathbf{E}Y)^2$, i.e. $\mathbf{V}Y(T) = 0$. □

It is clear that the controls μ^i cannot be constructed without knowing the function u. Nevertheless, the Theorem establishes the possibility, in principle, of arbitrarily reducing $\mathbf{V}Y(T)$ at the expense of choosing the functions μ^i in an appropriate manner.

We give a means of constructing μ^i reducing the variance. Suppose we know the solution of the problem (14.22), (14.23) with a replaced by a function $\tilde{a}$ close to it. Denote this solution by $\tilde{u}$. It can be understood that $\tilde{v} = \tilde{u}^2$ satisfies the equation (14.29) in which a is replaced by $\tilde{a}$. We write this equation as follows:

$$\min_{\mu}\left[\frac{\partial \tilde{v}}{\partial t} + \sum_{i=1}^{n} \mu^i \frac{\partial \tilde{v}}{\partial x^i} + \frac{1}{2}\sum_{i=1}^{n} \frac{\partial^2 \tilde{v}}{\partial x^{i2}} + \tilde{v}a + \sum_{i=1}^{n} \mu^{i2}\tilde{v} + \tilde{v}(\tilde{a} - a)\right] = 0.$$

This equation is clearly the Bellman equation in the problem of minimising the functional

$$I_\rho(\mu^1, \ldots, \mu^n) = \mathbf{E}\left(Z_{t,x,z}(T) + \int_t^T \rho(s, X_{t,x}(s)) Z_{t,x,z}(s)\, ds\right) \tag{14.30}$$

along the system (14.26), where $\rho = \tilde{v}(\tilde{a} - a)$, and $\tilde{v}z$ is the minimum of this functional. The function $\tilde{v}z$ is equal to the value of the righthand side of (14.30) along the system for

$$\mu^i = \tilde{\mu}^i = -\frac{1}{2\tilde{v}}\frac{\partial \tilde{v}}{\partial x^i}.$$

The variance $\mathbf{V}Y(T)$, computed along the system (14.26) with $\mu^i = \tilde{\mu}^i$, is equal to

$$\mathbf{V}Y(T) = \mathbf{E}Z(T) - (\mathbf{E}Y)^2 = \mathbf{E}Z(T) - v. \tag{14.31}$$

Since ρ is small, $\tilde{v}$ is close to v and $\mathbf{E}Z(T)$ is close to $\tilde{v}$. Therefore, by (14.31), $\mathbf{V}Y(T)$ is small.

REMARK 14.2. As shown in this Subsection, the introduction of the system (14.24) with suitable μ^i allows us to reduce the error in the Monte-Carlo method. At the same time it has a more general form in comparison with the system (14.2). To integrate it we can use the methods of order of accuracy two expounded in Chapter 3.

14.4. Examples of numerical experiments. The majority of numerical experiments presented in this Subsection are related with the computation of the *Wiener integral*

$$I = \int_{\mathbf{C}^n} \exp\left[\alpha \int_0^T x^2(s)\, ds\right] d_w(x). \tag{14.32}$$

The system (14.2) leads to the system of two equations

$$dX(s) = dw(s), \qquad X(t) = x,$$
$$dY(s) = \alpha Y(s) X^2(s)\, ds, \qquad Y(t) = y. \tag{14.33}$$

Recall that

$$I = \mathbf{E}Y_{0,0,1}(T) = \mathbf{E}\exp\left[\alpha \int_0^T X_{0,0}^2(s)\, ds\right], \tag{14.34}$$

where $X_{0,0}(s)$, $Y_{0,0,1}(s)$ is the solution of the system (14.33) for $t = 0$, $x = 0$, $y = 1$. As in the previous Subsection, we introduce the function

$$u(t,x) = \mathbf{E}Y_{t,x,1}(T) = \mathbf{E}\exp\left[\alpha\int_t^T X_{t,x}^2(s)\,ds\right]. \tag{14.35}$$

This function satisfies the equation

$$\frac{\partial u}{\partial t} + \frac{1}{2}\frac{\partial^2 u}{\partial x^2} + \alpha x^2 u = 0 \tag{14.36}$$

and initial condition

$$u(T,x) = 1. \tag{14.37}$$

The solution of (14.36), (14.37) for $\alpha = \lambda^2/2$, $0 \le \lambda < \pi/(2T)$, has the form

$$u(t,x) = \exp\left[\frac{1}{2}\lambda x^2 \tan(\lambda(T-t)) - \frac{1}{2}\ln\cos(\lambda(T-t))\right]. \tag{14.38}$$

Since $I = u(0,0)$ (see (14.34), (14.35)), (14.38) leads to the well-known result

$$I = (\cos\lambda T)^{-1/2}, \qquad \alpha = \frac{\lambda^2}{2}, \qquad 0 \le \lambda < \frac{\pi}{2T}. \tag{14.39}$$

Further, the solution of (14.36), (14.37) for $\alpha = -\lambda^2/2$ has the form

$$\begin{aligned} u(t,x) = \exp\Bigg[&\frac{\frac{1}{2}\lambda\,(1-\exp[2\lambda(T-t)])}{1+\exp[2\lambda(T-t)]}x^2 + \frac{\lambda(T-t)}{2} \\ &+ \frac{1}{2}\ln\frac{2}{1+\exp[2\lambda(T-t)]}\Bigg], \qquad \alpha = -\frac{\lambda^2}{2}, \end{aligned} \tag{14.40}$$

and, consequently,

$$I = \left[\frac{2\exp(\lambda T)}{1+\exp(2\lambda T)}\right]^{1/2}, \qquad \alpha = -\frac{\lambda^2}{2}. \tag{14.41}$$

We look for the variance of the variable $Y_{0,0,1}$. For $\alpha = \lambda^2/2$:

$$\begin{aligned} \mathbf{V}Y_{0,0,1}(T) &= \mathbf{E}Y_{0,0,1}^2(T) - (\mathbf{E}Y_{0,0,1}(T))^2 \\ &= \mathbf{E}\exp\left(2\alpha\int_0^T X_{0,0}^2(s)\,ds\right) - [\cos(\lambda T)]^{-1} \\ &= \left[\cos\left(\sqrt{2}\lambda T\right)\right]^{-1/2} - [\cos(\lambda T)]^{-1}, \qquad 0 \le \lambda < \frac{\pi}{2\sqrt{2}T}. \end{aligned} \tag{14.42}$$

For $\pi/(2\sqrt{2}T) \le \lambda \le \pi/(2T)$ the variance $\mathbf{V}Y_{0,0,1}(T)$ is equal to ∞. The same holds for $\alpha = -\lambda^2/2$:

$$\mathbf{V}Y_{0,0,1}(T) = \left[\frac{2\exp\left(\sqrt{2}\lambda T\right)}{1+\exp\left(2\sqrt{2}\lambda T\right)}\right]^{1/2} - \frac{2\exp(\lambda T)}{1+\exp(2\lambda T)}. \tag{14.43}$$

By Theorem 14.1, the variable $Y_{0,0,1}(T)$, computed along the system

$$\begin{aligned} dX &= \lambda \tan\left(\lambda(T-s)\right) X + dw(s), \qquad X(0) = 0, \\ dY &= \frac{\lambda^2}{2} X^2 Y\, ds - \lambda \tan\left(\lambda(T-s)\right) XY\, dw(s), \qquad Y(0) = 1, \end{aligned} \tag{14.44}$$

is deterministic for $\alpha = \lambda^2/2$. For $\alpha = -\lambda^2/2$ the variable $Y_{0,0,1}(T)$ becomes deterministic as the solution of the following system:

$$\begin{aligned} dX &= \frac{\lambda\left(1-\exp\left(2\lambda(T-s)\right)\right)}{1+\exp\left(2\lambda(T-s)\right)} X + dw(s), \qquad X(0) = 0, \\ dY &= -\frac{\lambda^2}{2} X^2 Y\, ds - \frac{\lambda\left(1-\exp\left(2\lambda(T-s)\right)\right)}{1+\exp\left(2\lambda(T-s)\right)} XY\, dw(s), \qquad Y(0) = 1. \end{aligned} \tag{14.45}$$

For completeness of exposition we give the derivation of, e.g., (14.40). To this end we change variables in (14.36) for $\alpha = -\lambda^2/2$:

$$u = \exp v.$$

The relations (14.36), (14.37) give

$$\frac{\partial v}{\partial t} + \frac{1}{2}\left(\frac{\partial v}{\partial x}\right)^2 + \frac{1}{2}\frac{\partial^2 v}{\partial x^2} - \frac{\lambda^2}{2}x^2 = 0, \qquad v(T,x) = 0. \tag{14.46}$$

We will look for a solution of the problem (14.46) in the form

$$v(t,x) = \frac{1}{2}p(t)x^2 + r(t), \qquad p(T) = r(T) = 0.$$

For $p(t)$ and $r(t)$ we obtain the Cauchy problem

$$\begin{aligned} p' + p^2 - \lambda^2 &= 0, \qquad p(T) = 0, \\ r' + \frac{1}{2}p &= 0, \qquad r(T) = 0, \end{aligned}$$

whose solution leads to formula (14.40).

Table 14.1: Computation of a Wiener integral by methods from §10.

α	h	N	$\mathbf{E}Y(1)$	M_I	M_{II}	M_{III}	$\frac{2}{\sqrt{N}}[\mathbf{V}Y(1)]^{1/2}$
	0.2	100	0.6776	0.6576 ± 0.0547	0.6316 ± 0.0494	0.6330 ± 0.0494	0.0475
	0.1	100	0.6776	0.6810 ± 0.0532	0.6688 ± 0.0510	0.6695 ± 0.0510	0.0475
	0.01	100	0.6776	0.6650 ± 0.0442	0.6634 ± 0.0441	0.6640 ± 0.0441	0.0475
-1	0.2	10 000	0.6776	0.6955 ± 0.0052	0.6713 ± 0.0049	0.6749 ± 0.0048	0.0048
	0.1	10 000	0.6776	0.6841 ± 0.0049	0.6743 ± 0.0048	0.6749 ± 0.0048	0.0048
	0.2	100 000	0.6776	0.6973 ± 0.0016	0.6733 ± 0.0015	0.6769 ± 0.0015	0.0015
	0.2	100	0.8050	0.8008 ± 0.0376	0.7734 ± 0.0369	0.7749 ± 0.0365	0.0344
	0.1	100	0.8050	0.8086 ± 0.0371	0.7961 ± 0.0375	0.7967 ± 0.0373	0.0344
-0.5	0.01	100	0.8050	0.8018 ± 0.0302	0.8007 ± 0.0329	0.8007 ± 0.0303	0.0344
	0.2	10 000	0.8050	0.8254 ± 0.0034	0.8011 ± 0.0035	0.8030 ± 0.0035	0.0034
	0.1	10 000	0.8050	0.8135 ± 0.0034	0.8028 ± 0.0035	0.8032 ± 0.0034	0.0034
	0.2	100	1.3604	1.2759 ± 0.0763	1.4222 ± 0.1455	1.4443 ± 0.1663	0.1651
	0.1	100	1.3604	1.3093 ± 0.0971	1.3928 ± 0.1352	1.3999 ± 0.1397	0.1651
0.5	0.01	100	1.3604	1.3048 ± 0.0654	1.3105 ± 0.1342	1.3111 ± 0.1344	0.1651
	0.2	10 000	1.3604	1.2356 ± 0.0068	1.3453 ± 0.0124	1.3598 ± 0.0143	0.0165
	0.1	10 000	1.3604	1.2865 ± 0.0087	1.3524 ± 0.0126	1.3572 ± 0.0132	0.0165

EXAMPLE 14.1. In Table 14.1 we have given the results of integrating the system (14.33) over the interval $[0, 1]$ with initial data $X(0) = 0$, $Y(0) = 1$ for the α's indicated, by the methods of first (M_I), second (M_{II}), and third (M_{III}) orders of accuracy constructed in §10 for systems with additive noises. The system (14.33) is a system with a single noise. Therefore the methods of first (Euler's method) and second (see (10.2)) orders of accuracy require the modeling of one random variable at each step. In our case, the method of third order of accuracy (see (10.36)) requires the modeling of two random variables at each step. This is related to the fact that i and r in (10.36) can take only the value one (since $q = 1$) and ζ_{1k}^2 can therefore be replaced by one (see (10.35)). To obtain the results, the variable ξ has (in all three methods) been modeled by the $N(0,1)$-distribution, and the variable ν has been modeled by the law (see (10.35)) $\mathbf{P}(\nu = -1/\sqrt{12}) = \mathbf{P}(\nu = 1/\sqrt{12}) = 1/2$.

The numbers presented in Table 14.1 are an approximation for $\mathbf{E}\overline{Y}(1)$, as computed by

$$\mathbf{E}\overline{Y}(1) \doteq \frac{1}{N}\sum_{m=1}^{N} \overline{Y}^{(m)}(1) \pm \frac{2}{N^{1/2}}\left\{\frac{1}{N}\sum_{m=1}^{N}\left[\overline{Y}^{(m)}(1)\right]^2 - \left[\frac{1}{N}\sum_{m=1}^{N}\overline{Y}^{(m)}(1)\right]^2\right\}^{1/2}, \tag{14.47}$$

i.e. under the assumption that the sample variance is sufficiently close to $\mathbf{V}\overline{Y}(1)$, the quantity $\mathbf{E}\overline{Y}(1)$ lies between the given limits with probability 0.95. It is obvious that the true value of the required Wiener integral (14.32) at $T = 1$, which is equal to $\mathbf{E}Y(1)$, differs from $\mathbf{E}\overline{Y}(1)$ by $O(h)$ for Euler's method, by $O(h^2)$ for the method (10.2), and by $O(h^3)$ for the method (10.36). It is also obvious that with increasing N the error of the Monte-Carlo method reduces, while as $N \to \infty$, the difference between the tabulated values and the true value of $\mathbf{E}Y(1)$ tends to the error of the numerical integration. For $\alpha = -1$ and $h = 0.1, 0.2$ it can be seen from the Table that for Euler's method this error is one-two units of the second position after the period, for the method of second order it is several units of the third position while for the method of third order it is even less. As α increases, the efficiency of the methods of higher order becomes even more transparent.

There is a column for $(2/\sqrt{N})[\mathbf{V}Y(T)]^{1/2}$ in the Table. This column differs from the sample values of $(2/\sqrt{N})[\mathbf{V}\overline{Y}(T)]^{1/2}$, of the component

$$\frac{2}{\sqrt{N}}\left[\frac{1}{N}\sum_{m=1}^{N}\left(\overline{Y}^{(m)}(1)\right)^2-\left(\frac{1}{N}\sum_{m=1}^{N}\overline{Y}^{(m)}(1)\right)^2\right]^{1/2},$$

first because of the estimation error, and secondly because of the numerical integration error. By (14.19), the latter is $(2/\sqrt{N})O(h)$. By themselves, the components in the columns M_I, M_{II}, and M_{III} cannot become smaller as the order of accuracy increases. With an increase of the order of accuracy these components can become close to $(2/\sqrt{N})\left(\mathbf{V}Y(T)\right)^{1/2}$, because the numerical integration error decreases. Moreover, they may also increase (see, e.g., the data for $\alpha = 0.5$). We stress once more that the numbers listed in the Table include the numerical integration error, and therefore the indicated region of variation of them need not cover the true value, especially for methods of low order (see, e.g., the data M_I for $h = 0.2$, $N = 10\,000$).

Table 14.2: Computation of a Wiener integral by methods from §14.

α	h	N	$\mathbf{E}Y(1)$	M_E	M_{R-K}	$\frac{2}{\sqrt{N}}[\mathbf{V}Y(1)]^{1/2}$
	0.2	100	0.6776	0.7299 ± 0.0441	0.7025 ± 0.0436	0.0475
	0.2	10 000	0.6776	0.6987 ± 0.0051	0.6770 ± 0.0048	0.0048
−1	0.1	100	0.6776	0.7035 ± 0.0483	0.6883 ± 0.0477	0.0475
	0.1	10 000	0.6776	0.6876 ± 0.0048	0.6779 ± 0.0047	0.0048
	0.2	100	0.8050	0.8502 ± 0.0273	0.8253 ± 0.0295	0.0344
	0.2	10 000	0.8050	0.8279 ± 0.0034	0.8050 ± 0.0034	0.0034
−0.5	0.1	100	0.8050	0.8259 ± 0.0231	0.8122 ± 0.0339	0.0344
	0.1	10 000	0.8050	0.8160 ± 0.0034	0.8055 ± 0.0034	0.0034
	0.2	100	1.3604	1.1864 ± 0.0429	1.2685 ± 0.0688	0.1651
	0.2	10 000	1.3604	1.2297 ± 0.0065	1.3536 ± 0.0151	0.0165
0.5	0.1	100	1.3604	1.2611 ± 0.0762	1.3322 ± 0.1126	0.1651
	0.1	10 000	1.3604	1.2824 ± 0.0089	1.3540 ± 0.0134	0.0165

EXAMPLE 14.2. In Table 14.2 we have given the results of integrating the system (14.33) by Euler's method (M_E) and the Runge–Kutta method of fourth order of accuracy (M_{R-K}). It can be seen from this Table that, e.g., for $\alpha = 0.5$ the numerical integration error in Euler's method is more than one unit of the first position after the period, while in the Runge–Kutta method it moves to the unit of the third position. We draw attention to the fact that the sample variance in the three last columns of the Table is less for the Euler method than for the Runge–Kutta method. As already noted in the previous Example, the 'large' variance of the Runge–Kutta method can in this case be explained by its greater accuracy.

EXAMPLE 14.3. For relatively small ϵ we compute the integral

$$I(\lambda,\epsilon)=\int_{\mathbf{C}}\exp\left(\frac{\lambda^2}{2}\int_0^T\left[x^2(s)+\epsilon g(x(s))\right]ds\right)d_wx.$$

To reduce the *error of the Monte-Carlo method*, in accordance with §14.3 we may take for $\tilde{a}$ the function $\tilde{a} = \lambda^2 x^2/2$. Then we have to integrate the system (see (14.44))

$$\begin{aligned} dX &= p(s)X\,ds + dw(s), \qquad X(0) = 0, \\ dY &= Y\frac{\lambda^2}{2}\left(X^2 + \epsilon g(X)\right) ds - p(s)XY\,dw(s), \qquad Y(0) = 1, \end{aligned} \tag{14.48}$$

where $p(s) = \lambda \tan(\lambda(T - s))$.

We give the results of the approximate integration by Euler's method for the system (14.48) and the system

$$\begin{aligned} dX &= dw(s), \qquad X(0) = 0, \\ dY &= Y\frac{\lambda^2}{2}\left(X^2 + \epsilon g(X)\right) ds, \qquad Y(0) = 1, \end{aligned} \tag{14.49}$$

for $\lambda = 1$, $T = 1$, $\epsilon g(x) = 0.1 \cos x \cdot x^2$, $h = 0.0005$, $N = 100$:

for (14.48) we have $\mathbf{E}\overline{Y}(1) = 1.3599 \pm 0.0053$;
for (14.49) we have $\mathbf{E}\overline{Y}(1) = 1.3747 \pm 0.0630$.

From this it is clear that the error of the Monte-Carlo method for the system (14.48) is approximately 10 times smaller than the error of the Monte-Carlo method for the system (14.49).

Table 14.3: Computation of a multidimensional Wiener integral.

h	N	M_E	M_{R-K}
0.2	25	0.1142 ± 0.0790	0.1410 ± 0.0536
0.2	100	0.0616 ± 0.0503	0.1230 ± 0.0231
0.2	1000	0.0848 ± 0.0179	0.1785 ± 0.0711
0.1	100	0.0997 ± 0.0262	0.1094 ± 0.0239
0.1	1000	0.1126 ± 0.0089	0.1220 ± 0.0081
0.1	10 000	0.1166 ± 0.0029	0.1270 ± 0.0026
0.05	10 000	0.1215 ± 0.0027	0.1264 ± 0.0026

EXAMPLE 14.4. By the *Cameron–Martin formula* (see [23, p. 323]), the value of the Wiener integral of the functional

$$V(x(\cdot)) = \exp\left(-\int_0^1 \left(x_1^2 + 2x_2^2 + 2x_3^2 + x_4^2 + x_1x_2 + x_2x_3 + x_3x_4\right) ds\right)$$

is equal to

$$\exp\left(\frac{1}{2}\int_0^1 \operatorname{tr}\Gamma(s)\,ds\right),$$

where the matrix $\Gamma(s)$ can be found from the Cauchy problem

$$\frac{d\Gamma(s)}{ds} = 2Q - \Gamma^2(s), \qquad Q = \begin{pmatrix} 1 & 1/2 & 0 & 0 \\ 1/2 & 2 & 1/2 & 0 \\ 0 & 1/2 & 2 & 1/2 \\ 0 & 0 & 1/2 & 1 \end{pmatrix}, \qquad \Gamma(1) = 0.$$

Numerically integrating this system gives the above-mentioned Wiener integral: $I \doteq 0.1285$.

Computations of this integral via integration of the system (14.2) by the Euler and Runge–Kutta methods are presented in Table 14.3. As in Table 14.1 and Table 14.2, next to the error of the Monte-Carlo method listed there arises another error, $O(h)$ for the Euler method and $O(h^4)$ for the Runge–Kutta method. As can be seen from the table, this error appears in an essential way in the Euler method (see, e.g., the last row of Table 14.3, where the exact value 0.1285 of this integral is not covered by the values 0.1215 ± 0.0027), and does not appear in the Runge–Kutta method when taking the amount of trajectories. Recall that for the problems under consideration the amount of computations in the Runge–Kutta method for given h and N is only twice as much as in the Euler method.

Table 14.4: Comparison of various methods of numerical modeling.

t_i	0.1	0.2	0.3	0.4	0.5	0.6	0.7	0.8	0.9	1.0
$\hat{x}(t_i)$	1.11	1.01	0.73	0.43	0.70	0.47	0.50	0.52	0.28	0.30
$\hat{x}_i$	1.15	1.09	0.68	0.34	0.65	0.42	0.48	0.48	0.25	0.34
$\overline{x}_i$	0.80	1.20	0.26	0.28	0.98	0.05	0.82	0.30	0.15	0.50

EXAMPLE 14.5. Consider the stationary scalar system

$$\begin{aligned} dx &= ax\,dt + b\,dw(t), \qquad & x(0) &= 0, \\ dz &= cx\,dt + d\,dv(t), \qquad & z(0) &= 0, \end{aligned} \tag{14.50}$$

where $w(t)$ and $v(t)$ are uncorrelated standard Wiener processes.

The discretisation step for modeling $\hat{x}(t)$ by (4.3)–(4.5) can be taken sufficiently small such that the values of $\hat{x}(t_i)$ listed in Table 14.4 differ but little from the exact values. The value $\hat{x}_i$ can be modeled by (4.19)–(4.21), and the value $\overline{x}_i$ by (4.7). The discretisation step for modeling $\hat{x}_i$ and $\overline{x}_i$ has been taken equal to $h = 0.1$; $a = -2$, $b = 0.8$, $c = 2$, $d = 0.1$, $x_0 = 1$.

EXAMPLE 14.6. In the numerical modeling of the *Kalman–Bucy filter* great importance is attached to the dependence of accuracy on the size of the discretisation step. Of course, we can make $\hat{x}_i$ arbitrarily close to $\hat{x}(t_i)$ by choosing a sufficiently small discretisation step h. The error of the approximation of $\hat{x}_i$ to $x(t_i)$ has two components. First, there is the proper filtering error, i.e. the deviation of $\hat{x}(t_i)$ from $x(t_i)$, and secondly there is the discretisation error, i.e. the deviation of $\hat{x}_i$ from $\hat{x}(t_i)$. It is clear that it does not make sense to make the approximation of $\hat{x}_i$ to $\hat{x}(t_i)$ more precise than the estimation of $x(t_i)$ by $\hat{x}(t_i)$. In other words, when choosing the discretisation step we have to estimate the proper filtering error and choose h such that the discretisation error is comparable to it.

These considerations can be used in practice in various ways. For example, in the scalar case we may compare $p(t_i) = \mathbf{E}\left(x(t_i) - \hat{x}(t_i)\right)^2$ and $p_i = \mathbf{E}\left(x(t_i) - \hat{x}_i\right)^2$. The computation of these quantities can be done beforehand for various h. Clearly, $p_i > p(t_i)$ and p_i tends to $p(t_i)$ as $h \to 0$. We will choose h as large as possible, but such that the inequality

$$p_i \leq (1+\epsilon)p(t_i), \qquad \epsilon > 0,$$

holds. In a similar way we can choose a suitable numerical modeling step when using the Euler method. In that case the inequality

$$\overline{p}_i = \mathbf{E}\left(x(t_i) - \overline{x}_i\right)^2 \leq (1+\epsilon)p(t_i)$$

must hold.

We compute the values of $p(t_i)$ and p_i using (4.5) and (4.21). We will obtain formulas for computing the error $\overline{p}_i$ of the Euler method.

Equation (4.7) for the estimator $\overline{x}_i$ takes the following form for the system (14.50) (we assume that $h_i = h$):

$$\overline{x}_{i+1} = \overline{x}_i + ah\overline{x}_i + \frac{c}{d^2}\left(z(t_i+1) - z(t_i) - ch\overline{x}_i\right)p(t_i), \qquad \overline{x}_0 = m_0.$$

Using (4.11), (4.12) we obtain

$$\overline{x}_{i+1} = \left(1 + ah - \frac{c^2}{d^2}p(t_i)h\right)\overline{x}_i + \frac{c}{d^2}p(t_i)g_i x(t_i) + \frac{c}{d^2}p(t_i)v_i. \tag{14.51}$$

We put

$$\overline{p}_{i+1} = \mathbf{E}\left(x(t_{i+1}) - \overline{x}_{i+1}\right)^2 = \mathbf{E}x^2(t_{i+1}) + \mathbf{E}\overline{x}_{i+1}^2 - 2\mathbf{E}x(t_{i+1})\overline{x}_{i+1}.$$

Using Theorem 4.1 and (14.51), it is not difficult to find recurrence relations for $\mathbf{E}x^2(t_{i+1})$, $\mathbf{E}\overline{x}_{i+1}^2$, and $\mathbf{E}x(t_{i+1})\overline{x}_{i+1}$:

$$\mathbf{E}x^2(t_{i+1}) = e^{2ah}\mathbf{E}x^2(t_i) + q_i(t_{i+1}),$$

$$\begin{aligned}\mathbf{E}x(t_{i+1})\overline{x}_{i+1} &= e^{ah}\left(1 + ah - \frac{c^2}{d^2}p(t_i)h\right)\mathbf{E}x(t_i)\overline{x}_i \\ &\quad + e^{ah}\frac{c}{d^2}p(t_i)g_i\mathbf{E}x^2(t_i) + \frac{c}{d^2}p(t_i)s_i(t_{i+1}),\end{aligned}$$

$$\begin{aligned}\mathbf{E}\overline{x}_{i+1}^2 &= \left(1 + ah - \frac{c^2}{d^2}p(t_i)h\right)^2\mathbf{E}\overline{x}_i^2 + \left(\frac{c}{d^2}p(t_i)g_i\right)^2\mathbf{E}x^2(t_i) \\ &\quad + 2\left(1 + ah - \frac{c}{d^2}p(t_i)h\right)\frac{c}{d^2}p(t_i)g_i\mathbf{E}x(t_i)\overline{x}_i + \left(\frac{c}{d^2}p(t_i)g_i\right)^2 r_i(t_{i+1}).\end{aligned}$$

The quantities $g_i(t_{i+1})$, $s_i(t_{i+1})$, $r_i(t_{i+1})$ can be found from (4.14) for $h = h_i$, and

$$g_i = \frac{c}{d}(e^{ah} - 1).$$

We have to note that in Euler's method $\overline{p}_i$ depends on x_0 , while p_i does not depend on x_0; this is an additional advantage of the optimal filtering method with the use of information at discrete times. In Table 14.5 we have given the largest possible discretisation steps for the given values of ϵ; h_1 is the step in Euler's method with $x_0 = 10$; h_2 is the step in Euler's method with $x_0 = 1$; and h is the step of the method (4.19)–(4.21). The computations have been done for the system (14.50) with the given values of the parameters a, b, c, d.

The results presented in Table 14.5 are sufficiently typical. They, as well as other numerical experiments, allow one to conclude that the discretisation step in the method (4.19)–(4.21) can be chosen 2–6 times larger than in Euler's method (see [31] for additional details).

Table 14.5: Choice of the discretisation step.

$a=-2,\quad b=2,\quad c=1,\quad d=0.1$						
ϵ	0.01	0.03	0.05	0.1	0.3	0.5
h_1	0.005	0.009	0.011	0.015	0.027	0.035
h_2	0.008	0.015	0.019	0.028	0.050	0.055
h	0.017	0.032	0.043	0.068	0.155	0.213
$a=-1,\quad b=0.1,\quad c=10,\quad d=0.1$						
h_1	0.001	0.001	0.001	0.002	0.004	0.005
h_2	0.007	0.013	0.017	0.024	0.040	0.050
h	0.035	0.065	0.087	0.137	0.310	0.427

Bibliography

[1] T.A. Averina, S.S. Artem'ev, A new family of numerical methods for solving stochastic differential equations, *Dokl. Akad. Nauk. SSSR* **288** No. 4 (1986), pp. 777–780. (In Russian.)

[2] I.S. Berezin, N.P. Zhidkov, *Computing methods*, Fizmatgiz, Moscow, 1962. (In Russian.)

[3] A.D. Wentsell, *A course in the theory of random processes*, Nauka, Moscow, 1975. (In Russian.)

[4] A.D. Wentsell, S.A. Gladyshev, G.N. Mil'steĭn, Piecewise constant approximation for Monte-Carlo computation of Wiener integrals, *Teor. Veroyatn. i Primen.* **29** No. 4 (1984), pp. 715–722. (In Russian.)

[5] R.E.A.C. Paley, N. Wiener, *Fourier transforms in the complex domain*, AMS, New York, (1934),

[6] I.M. Gel'fand, A.M. Yaglom, Integration in function spaces and its application in quantum physics, *Uspekhi Mat. Nauk.* **11** No. 1 (1956), pp. 77-114. (In Russian.)

[7] I.I. Gikhman, A.V. Skorokhod, *Introduction to the theory of random processes*, Fizmatgiz, Moscow, 1965. (In Russian.)

[8] I.I. Gikhman, A.V. Skorokhod, *Stochastic differential equations*, Nauk. Dumka, Kiev, 1967. (In Russian.)

[9] I.I. Gikhman, A.V. Skorokhod, *Controllable stochastic processes*, Nauk. Dumka, Kiev, 1977. (In Russian.)

[10] I.I. Gikhman, A.V. Skorokhod, *Stochastic differential equations and their applications*, Nauk. Dumka, Kiev, 1982. (In Russian.)

[11] S.A. Gladyshev, G.N. Mil'steĭn, A Runge–Kutta method for computing Wiener integrals of functionals of exponential type, *Zh. Vychisl. Mat. i Mat. Fiz.* **24** No. 8 (1984), pp. 1136–1150. (In Russian.)

[12] E.B. Dynkin, *Markov processes*, Fizmatgiz, Moscow, 1963. (In Russian.)

[13] A.D. Egorov, P.I. Sobolevskiĭ, D.A. Yanovich, *Approximate methods for computing continual integrals*, Nauka i Tekhn., Minsk, 1985. (In Russian.)

[14] B.S. Elepov, A.A. Kronberg, G.A. Mikhaĭlov, K.K. Sabel'fel'd, *Solution of boundray value problems by the Monte-Carlo method*, Nauka, Novosibirsk, 1980. (In Russian.)

[15] S.M. Ermakov, *The Monte-Carlo method and related questions*, Nauka, Moscow, 1971. (In Russian.)

[16] S.M. Ermakov, G.A. Mikhaĭlov, *A course of statistical modeling*, Nauka, Moscow, 1976. (In Russian.)

[17] S.M. Ermakov, V.V. Nekrutkin, A.S. Sipin, *Random processes for solving the equations of mathematical physics*, Nauka, Moscow, 1984. (In Russian.)

[18] K. Itô, H.P. McKean, Jr., *Diffusion processes and their sample paths*, Springer, 1974.

[19] R.E. Kalman, P.L. Falb, M.A. Arbib, *Topics in mathematical control theory*, McGraw-Hill, New York, 1969.

[20] P.I. Kitsul, On continuously-discrete filtering of Markov processes of diffusion type, *Avtomatik. i Telemekhanik.* **11** (1970), pp. 29–37. (In Russian.)

[21] N.V. Krylov, *Controllable processes of diffusion type*, Nauka, Moscow, 1977. (In Russian.)

[22] H.J. Kushner, *Probability methods for approximations in stochastic control and for elliptic equations*, Acad. Press, New York, 1977.

[23] P.S. Lipster, A.N. Shiryaev, *Statistics of random processes*, Nauka, Moscow, 1974. (In Russian.)

[24] H.P. McKean, Jr., *Stochastic calculus*, Acad. Press, New York, 1969.

[25] G.N. Mil'steĭn, Approximate integration of stochastic differential equations, *Teor. Veroyatn. i Prilozhen.* **19** No. 3 (1974), pp. 583–588. (In Russian.)

[26] G.N. Mil'steĭn, A method of second order of accuracy for integrating stochastic differential equations, *Teor. Veroyatn. i Prilozhen.* **23** No. 2 (1978), pp. 414–419. (In Russian.)

[27] G.N. Mil'steĭn, Probabilistic solution of linear systems of elliptic and parabolic equations, *Teor. Veroyatn. i Prilozhen.* **23** No. 4 (1978), pp. 851–855. (In Russian.)

[28] G.N. Mil'steĭn, Weak approximation of solutions of systems of stochastic differential equations, *Teor. Veroyatn. i Prilozhen.* **30** No. 4 (1985), pp. 706–721. (In Russian.)

[29] G.N. Mil'steĭn, Theorem on the order of convergence of mean-square approximations of solutions of systems of stochastic differential equations, *Teor. Veroyatn. i Prilozhen.* **32** No. 4 (1987), pp. 809–811. (In Russian.)

[30] G.N. Mil'steĭn, *Numerical integration of stochastic differential equations*, Izd. Ural. Univ., Sverdlovsk, 1988. (In Russian.)

[31] G.N. Mil'steĭn, S.A. P'yanzin, Numerical modeling of the Kalman–Bucy filter and optimal filters for discrete arrival of information, *Avtomatik. i Telemekhanik.* **46** No. 1 (1985), pp. 59–68. (In Russian.)

[32] G.N. Mil'steĭn, S.A. P'yanzin, Regularisation and numerical modeling of the Kalman–Bucy filter for systems with degenerate noises in the observations, *Avtomatik. i Telemekhanik.* **48** No. 11 (1987), pp. 80–92. (In Russian.)

[33] N.N. Nikitin, S.V. Pervachev, V.D. Razevig, On computer solution of stochastic differential equations of servomechanisms, *Avtomatik. i Telemekhanik.* **36** No. 4 (1975), pp. 133–137. (In Russian.)

[34] N.N. Nikitin, V.D. Razevig, Methods of numerical modeling of stochastic differential equations and estimates of their error, *Zh. Vychisl. Mat. i Mat. Fiz.* **18** No. 1 (1978), pp. 106–117. (In Russian.)

[35] K. Ostrem, *Introduction to the stochastic theory of equations*, Mir, Moscow, 1973. (In Russian.)

[36] E. Platen, An approximation method for a class of Itô processes, *Lit. Mat. Sb.* **21** No. 1 (1981), pp. 121–133. (In Russian.)

[37] V.S. Pugachev, I.N. Sinitsyn, Stochastic differential equations, Nauka, Moscow, 1985. (In Russian.)

[38] V.D. Razevig, Numerical modeling of multidimensional dynamical systems under random perturbations, *Avtomatik. i Telemekhanik.* **41** No. 4 (1980), pp. 177–186. (In Russian.)

[39] Yu.V. Rakitskiĭ, S.M. Ustinov, I.G. Chernorutskiĭ, *Numerical methods for solving stiff systems*, Nauka, Moscow, 1979. (In Russian.)

[40] W.H. Fleming, R.W. Rishel, *Deterministic and stochastic optimal control*, Springer, 1975.

[41] R.Z. Khas'minskiĭ, *Stability of systems of differential equations under random perturbations of their parameters*, Nauka, Moscow, 1969. (In Russian.)

[42] E. Hille, R.S Phillips, *Functional analysis and semigroups*, AMS, Providence, 1957.

[43] G. Hall, J.M. Watt, *Modern numerical methods for ordinary differential equations*, Clarendon Press, Oxford, 1976.

[44] H.J. Stetter, *Analysis of discretization methods for ordinary differential equations*, Springer, 1973.

[45] L.A. Yanovich, *Approximate computation of continual integrals with respect to Gaussian measures*, Nauka i Tekhn., Minsk, 1976.

[46] A.J. Chorin, Accurate evaluation of Wiener integrals, *Math. Comput.* **27** No. 121 (1973), pp. 1–15.

[47] J.M.C. Clark, R.T. Cameron, The maximum rate of convergence of discrete approximations for stochastic differential equations, *Lect. notes Control & Inform. Sci.* **25** (1980), pp. 162–171.

[48] H.S. Greenside, E. Helfand, Numerical integration of stochastic differential equations, *Bell Syst. Techn. J.* **60** No. 8 (1981), pp. 1927–1940.

[49] C.J. Harris, Modelling, simulation and control of stochastic systems with application in wastewater treatment, *Intern. J. Systems Sci.* **8** No. 4 (1977), pp. 393–411.

[50] E. Helfand, Numerical integrationof stochastic differential equations, *Bell Syst. Techn. J.* **58** No. 10 (1979), pp. 2289–2299.

[51] J.R. Klauder, W.P. Petersen, Numerical integration of multiplicative-noise stochastic differential equations, *SIAM J. Numer. Anal.* **6** (1985), pp. 1153–1166.

[52] G. Marujama, Continuous Markov processes and stochastic equations, *Rend. Mat. Circ. Palermo, Ser. 2* **4** (1955), pp. 48–90.

[53] N.J. Newton, An asymptotically efficient difference formula for solving stochastic differential equations, *Stochastics* **19** (1986), pp. 175–206.

[54] E. Pardoux, D. Talay, Discretization and simulation of stochastic differential equations, *Acta Appl. Math.* **3** No. 1 (1985), pp. 23–47.

[55] N.J. Rao, J.D. Borwankar, D. Ramkrishna, Numerical solution of Itô integral equations, *SIAM J. Control* **12** No. 1 (1974), pp. 124–139.

[56] W. Rumelin, Numerical treatment of stochastic differential equations, *SIAM J. Numer. Anal.* **19** No. 3 (1982), pp. 604–613.

[57] D. Talay, Résolution trajectorillée et analyse numérique des équations différentielles stochastiques, *Stochastics* **9** No. 4 (1983), pp. 275–306.

[58] D. Talay, How to discretize stochastic differential equations, *Lect. notes in Math.* **972** (1983), pp. 276–292.

[59] D. Talay, Efficient numerical schemes for the approximation of expectations of functionals of the solution of a SDE, and applications, *Lect. notes Control & Inform. Sci.* **61** (1984), pp. 294–313.

[60] D. Talay, Discrétisation d'une équation différentielle stochastique et calcul approché d'espérances de fonctionelles de la solution, *M2AN: Modél Math. et Anal. Numér.* **20** No. 1 (1986), pp. 141–179.

[61] U. Tetzlaff, H.-U. Zschiesche, Näherungslösungen für Itô-differentialgleichungen mittels Taylorentwicklung für Halbgruppen von Operatoren, *Wiss. Z. Techn. Hochschule Leuna-Merseburg* **2** (1984), pp. 332–339.

[62] W. Wagner, E. Platen, *Approximation of Itô integral equations*, Preprint ZIMM, Akad. Wissenschaft. DDR, Berlin, 1978.

Index

Other *Mathematics and Its Applications* titles of interest:

W.L. Miranker: *Numerical Methods for Stiff Equations and Singular Perturbation Problems*. 1980, 220 pp. ISBN 90-277-1107-0

K. Rektorys: *The Method of Discretization in Time and Partial Differential Equations*. 1982, 470 pp. ISBN 90-277-1342-1

L. Ixary: *Numerical Methods and Differential Equations and Applications*. 1984, 360 pp. ISBN 90-277-1597-1

B.S. Razumikhin: *Physical Models and Equilibrium Methods in Programming and Economics*. 1984, 368 pp. ISBN 90-277-1644-7

A. Marciniak: *Numerical Solutions of the N-Body Problem*. 1985, 256 pp. ISBN 90-277-2058-4

Y. Cherruault: *Mathematical Modelling in Biomedicine*. 1986, 276 pp. ISBN 90-277-2149-1

C. Cuvelier, A. Segal and A.A. van Steenhoven: *Finite Element Methods and Navier-Stokes Equations*. 1986, 500 pp. ISBN 90-277-2148-3

A. Cuyt (ed.): *Nonlinear Numerical Methods and Rational Approximation*. 1988, 480 pp. ISBN 90-277-2669-8

L. Keviczky, M. Hilger and J. Kolostori: *Mathematics and Control Engineering of Grinding Technology. Ball Mill Grinding*. 1989, 188 pp. ISBN 0-7923-0051-3

N. Bakhvalov and G. Panasenko: *Homogenisation: Averaging Processes in Periodic Media. Mathematical Problems in the Mechanics of Composite Materials*. 1989, 404 pp. ISBN 0-7923-0049-1

R. Spigler (ed.): *Applied and Industrial Mathematics*. Venice-1, 1989. 1991, 388 pp. ISBN 0-7923-0521-3

C.A. Marinov and P. Neittaanmaki: *Mathematical Models in Electrical Circuits. Theory and Applications*. 1991, 160 pp. ISBN 0-7923-1155-8

Z. Zlatev: *Iterative Improvement of Direct Solutions of Large and Sparse Problems*. 1991, 328 pp. ISBN 0-7923-1154-X

M. Vajtersic: *Algorithms for Elliptic Problems*. 1992, 352 pp. ISBN 0-7923-1918-4

V. Kolmanovskii and A. Myshkis: *Applied Theory of Functional Differential Equations*. 1992, 223 pp. ISBN 0-7923-2013-1

A.D. Egorov, P.I. Sobolevsky and L.A. Yanovich: *Functional Integrals: Approximate Evaluation and Applications*. 1993, 426 pp. ISBN 0-7923-2193-6

G.I. Marchuk: *Adjoint Equations and Analysis of Complex Systems*. 1994 ISBN 0-7923-3013-7

Other *Mathematics and Its Applications* titles of interest:

A. Bakushinsky and A. Goncharsky: *Ill-Posed Problems: Theory and Applications.* 1994, 256 pp. ISBN 0-7923-3073-0

G.N. Milstein: *Numerical Integration of Stochastic Differential Equations.* 1995, 170 pp. ISBN 0-7923-3213-X